M

A SHEARWATER BOOK

HIGH TECH TRASH

HIGH TECH TRASH

Digital Devices, Hidden Toxics, and Human Health

Elizabeth Grossman

○ **ISLAND**PRESS / Shearwater Books

Washington • Covelo • London

A Shearwater Book
Published by Island Press

Copyright © 2006 Elizabeth Grossman

Shearwater Books is a trademark of The Center for Resource Economics.

Library of Congress Cataloging-in-Publication data.
Grossman, Elizabeth, 1957-
    Digital devices, hidden toxics, and human health /
    Elizabeth Grossman.
        p. cm.
    Includes bibliographical references and index.
    ISBN 1-55963-554-1 (cloth : alk. paper)
    1. Waste electronic apparatus and appliances. 2. Electronic apparatus and appliances--Environmental aspects. 3. Electronic apparatus and appliances--Health aspects. 4. Product life cycle. I. Title.
TD799.85.G76 2006
363.72'87—dc22

                                                        2006004549

British Cataloguing-in-Publication data available.

Printed on recycled, acid-free paper ♲
Design by Brian Barth
Manufactured in the United States of America

10 9 8 7 6 5 4 3 2 1

For Emily, Jane, Olivia, and Phil, and for my parents

Contents

Preface

In August 2004 I stood on the tussocky tundra banks of Imnaviat Creek looking out toward the Brooks Range on the North Slope of Alaska and watched a biologist set up his laptop—an hour's walk from the nearest electrical outlet—and measure the depth of the permafrost. Last spring I called my parents in New York on a crystal clear wireless connection from a small town in Lappland. There are now computers in the Himalayas, the Andes, and the Amazon, and cell phone use is booming in rural Africa. The virtues of remaining unplugged aside, there is hardly a place left on earth to which someone has not brought a computer or mobile phone, and even those who write all their letters longhand now have lives that depend on digital technology.

High-tech electronics have become virtually ubiquitous and have transformed the world in ways that benefit us all. But for most of the forty or more years since commercial semiconductor and computer manufacture began, we have paid relatively little attention to the environmental and health impacts of producing and disposing of the microchip-powered gadgets that enable the Digital Age.

High-tech electronics are the most complex mass-produced consumer products ever manufactured—a complexity that presents special challenges when it comes to dealing with this equipment at the end of its useful life. And because the production of high-tech electronics involves many toxic and hazardous materials—and takes place on a global scale—their environmental impacts are now being felt by communities from the Arctic to Australia, with poorer countries and communities receiving a disproportionate share of the burden. If not addressed comprehensively and with solutions that show we have learned from past mistakes, these problems risk undermining the ecological and economic sustainability of affected communities worldwide, whether in Silicon Valley, the American Rust Belt, or southern China.

High technology has given us lightning-speed computation, instant messaging, and libraries without books, yet in creating the equipment that makes all this possible we have also unleashed tons of chemicals into the environment with impacts far more pernicious than an e-mail in-box full of spam. How we choose to make high-tech products and how we take out the high-tech trash will affect the quality of life for everyone from California to Africa, from Greenland to Malaysia, for decades to come. Computers and cell phones can be replaced, but watersheds and human beings cannot have their hard drives wiped and operating systems reinstalled if something goes wrong.

After learning how high-tech manufacturing was compromising the quality of the water in the Willamette River, which flows two minutes' walk from my front door, I set out to explore what other effects Information Age technology might be having on human health and the environment—and what is being done to solve these problems and improve on past practices in ways that will ensure a safer, cleaner, and healthier future.

▌▐▌▌▐

To place high-tech electronics in an ecological context—and to explore their physical connection to the natural world—I wanted to see what goes into making machines like the computer on which I am writing and what happens to them when they are discarded. While researching this book, I have worn a lot of borrowed hard hats and safety glasses and have gone through several sets of disposable earplugs. I have toured Superfund sites, descended to the bottom of a mine, visited with people whose homes are contaminated with toxic vapor, peered through the glass into clean rooms of an enormous semiconductor plant, watched old electronics be mined for gold, and have seen literally tons of discarded and dismantled computer equipment.

I have spoken to dozens of scientists who are trying to discover how chemicals embedded in and used to make the appliances that sit on our desks have wound up in people and in the food we eat—and what this means for our health and that of our children. I have spoken to people who make silicon wafers, semiconductors, computers, and all sorts of other high-tech devices. I have interviewed elected officials, experts in solid waste, engineers, and a former prison inmate. I have also taken notes at hours and hours of presentations by high-tech industry professionals, electronics recyclers, scientists, policy makers, and environmental advocates—all of too many different nationalities to name—who are working to understand and solve the problems posed by high-tech trash. While these problems are far more complex than I imagined, there are solutions on the horizon to at least some of them, and—thanks to the dedicated work of environmental advocates along with industrial engineers, manufacturers and legislators—some changes in how high-tech electronics are designed, produced, and disposed of are already under way.

There are many people to whom thanks are due for their help in making this book possible. For support from the John D. and Catherine T. MacArthur Foundation, I am deeply grateful and honored. For his faith in this project, his dedication, buoyant enthusiasm, and acumen, enormous thanks and admiration to Jonathan Cobb at Island Press.

Among those I would like to thank for being so generous with their time and information are Linda Birnbaum, Sam Blackman, Heather Bowman; Apple Chan, Kevin May, Lai Yun, and their colleagues at Greenpeace China; Lara Cushing, Gopal Dayaneni, Bette Fishbein, Robert Hale, Amanda Hawes, Rebecca Hayes, Bryant Hilton, Jon Hinck, Ronald Hites, Wanda Hudak, Sego Johnson, Iza Kruszewska, Theo

Lehner, Donna Lupardo, Jim Lynch, Tom MacDonald, Tim Mohin, Robert Houghton, Bob Moser, Kim Nauer, Gary Niekerk, Anne Peters, Jim Puckett, Wayne Rifer, Jeff Ruch, Tim Rudnicki, Greg Sampson, Caisa Samuelsson, Tom Sawyer, Arnold Schecter, Robin Schneider, Byron Sher, Ted Smith, Leroy Smith, Alan and Donna Turnbull, Joanna Underwood, Sarah Westervelt, Rick White, and Eric Williams. Thanks also to staff members at Boliden, Intel, Kuusakoski, Metech International, Noranda Recycling, and Phelps Dodge for making my visits possible, and to the Nation Institute, *Orion*, *Yes!* magazine, and the Woods Hole Marine Biological Laboratory's science journalism fellowship program—as well as Jessica Heise, Julie van Pelt, and everyone at Island Press who made this book possible.

Special thanks to Jerry Powell, Jonathan Brinckman, Rick Brown, John Carey, Rebecca Clarren, Andy Kerr, and other friends and colleagues who provided access to information and technical expertise; to Pamela Brody-Heine, Jackie Dingfelder, Betty Patton, and Lori Stole for their camaraderie and devotion to what they do; to Ed Gargan for his extraordinary hospitality in Beijing; and to Peter Eisner and Bill McKibben for their enthusiasm. Thanks also to Bill Fox, Gilly Lyons, Robert Stubblefield, Margot and George Thompson, and to my parents, Alvin and Sari Grossman.

HIGH TECH TRASH

CHAPTER ONE

The Underside of High Tech

The rapidity of change and the speed with which new situations are created follow the impetuous and heedless pace of man rather than the deliberate pace of nature.[1]
—Rachel Carson, *Silent Spring,* 1962

If future generations are to remember us with gratitude rather than with sorrow, we must achieve more than just the miracles of technology. We must leave them a glimpse of the world as God really made it, not just as it looked after we got through with it.[2]
—President Lyndon B. Johnson, 1965

A harbor seal arches her back and dives, a graceful comma of brown on the steel blue water of San Francisco Bay. A school of herring darts through the saltwater off the coast of Holland. A polar bear settles down to sleep in a den carved out of Arctic ice. A whale cruises the depths of the North Sea and a chinook salmon noses her way into the Columbia River on her way home to spawn. In the Gulf of Mexico, a bottlenose dolphin leaps above the waves. A seagoing tern lays an egg. A mother in Sweden nurses her baby, as does a mother in Oakland, California. Tissue samples taken from these animals and from these women's breasts contain synthetic chemicals used to make the plastics used in computers, televisions, cell phones, and other electronics resist fire. Americans have the highest levels of these compounds in their blood of any people yet tested, and the same chemicals have been found in food purchased in grocery stores throughout the United States.

On the shores of the Lianjiang River in southern China, a woman squats in front of an open flame. In the pan she holds over the fire is a smoky stew of plastic and metal—a melting circuit board. With unprotected hands she plucks out the microchips. Another woman wields a hammer and cracks the glass of an old computer monitor to remove the copper yoke. The lead-laden glass screen is tossed onto a riverside pile. Nearby, a man sluices a pan of acid over a pile of computer chips, releasing a puff of toxic steam. When the vapor clears a small fleck of gold will emerge. Up and down the river-banks are enormous hillocks of plastic and metal, the discarded remains of electronic appliances—monitors, keyboards, wires, printers, cartridges, fax machines, motors, disks, and cell phones—that have all been exported here for inexpensive, labor-intensive recycling. A bare-legged child stands on one of the mounds, eating an apple. At night, thick black smoke rises from a mountain of burning wires. In the southern Chinese city of Guiyu—one of the places in Asia where this primitive recycling takes place—an estimated 80 percent of the city's 150,000 residents are engaged in processing the million or more tons of electronic waste that have been arriving there each year since the mid-1990s.[3]

Mines that stretch for miles across the Arizona desert, that tunnel deep under the boreal forests of northern Sweden, and others on nearly every continent produce ore and metals that end up in electronic gadgets on desktops, in pockets, purses, and briefcases, and pressed close to ears all around the world. In a region of the Democratic Republic of the Congo wracked by horrific civil war, farmers have left their land to work in lucra-tive but dangerous, landslide-prone coltan mines. Sales of this ore, which is used in the manufacture of cell phones and other devices, have helped finance that war as well as the fighting between Uganda and Rwanda in this mineral-rich region of Africa. Although they are mostly hidden, met-als make up over half the material in the world's 660 to 700 million com-puters. A typical desktop computer can contain nearly thirty pounds of metal, and metals are used in all electronics that contain semiconductors and circuit boards (which are themselves 30 to 50 percent metal)—from big plasma screen TVs to tiny cell phones. Extracted and refined at great cost, about 90 percent of the metal that goes into electronics eventually ends up in landfills, incinerators, or some other kind of dump.

Traffic on the highway that runs between San Francisco and San Jose is bumper to bumper. Haze rises from the vehicle-clogged road. Office plazas, strip malls, and housing developments stretch out against the backdrop of hills that frame the valley. Pooled beneath the communities of Santa Clara, Cupertino, and Mountain View, California—to name but a few—are thousands of gallons of poisonous volatile organic compounds left by the manufacture of semiconductors. California's Silicon Valley now has more toxic waste sites subject to cleanup requirements under the federal government's Superfund program than any other region of comparable size in the United States. In parts of Mountain View, the U.S. Environmental Protection Agency (EPA) has found in groundwater levels of trichloroethylene (TCE)—a solvent used in semiconductor production that the EPA recognizes as a carcinogen—that may be sixty-five times more toxic than previously thought.[4] Official estimates say it will take decades, if not a century or more, to complete the cleanup. Families in Endicott and other communities in Broome and Dutchess counties in upstate New York are grappling with the same problem, living above a groundwater plume contaminated for over twenty years with TCE and other solvents used in microchip manufacture.

In the high desert country of New Mexico, the ochre and mustard colored cliffs of the Sandia Mountains rise above the Rio Grande valley. Globe mallow and prickly pear sprout from the sandy soil. This is the third most arid state in the nation, and the past decade has been marked by drought. Yet one of the handful of semiconductor manufacturers located near Albuquerque has been using about four million gallons of water a day—over thirty times the water an average American household uses annually[5]—while sending large quantities of toxics into the local waste stream. Similar scenarios have emerged in other parts of the country where semiconductor manufacture has taken place—among them, the Texas hill country around Austin, the Boston area landscape that gave rise to the American Revolution, and the suburban sprawl that surrounds Phoenix. Residents of Endicott, New York, and Rio Rancho, New Mexico, have asked the Agency for Toxic Substance and Disease Registry (part of the U.S. Department of Health and Human Services) to assess the health impacts of hazardous air pollutants—including trichloroethylene, methanol, ethylene chloride, and several perfluorocarbons—emitted by high-tech manufacturers located in their communities.

Semiconductors come off the assembly line in numbers that dwarf other manufactured products, but because microchips are so tiny, we're less inclined to think about their environmental footprint. One of Intel's Pentium 4 chips is smaller than a pinky fingernail and the circuit lines on the company's new Itanium 2 chips are smaller than a virus—too small to reflect a beam of light.[6] Producing something of this complexity involves many steps, each of which uses numerous chemicals and other materials and a great deal of energy. Research undertaken by scientists at United Nations University and the National Science Foundation found that at least sixteen hundred grams of fossil fuel and chemicals are needed to produce one two-gram microchip. Further, the secondary material used to produce such a chip amounts to 630 times the mass of the final product, a proportion far larger than for traditional low-tech items.[7] In 2004 some 433 billion semiconductors were produced worldwide.[8]

The Information Age. Cyberspace. The images are clean and lean. They offer a vision of business streamlined by smart machines and high-speed telecommunications and suggest that the proliferation of e-commerce and dot-coms will make the belching smokestacks, filthy effluent, and slag heaps of the Industrial Revolution relics of the past. With this in mind communities everywhere have welcomed high technology under the banner of "clean industry," and as an alternative to traditional manufacturing and traditional exploitation of natural resources. But the high-tech industry is far from clean.

Sitting at my desk in Portland, Oregon, the tap of a few keys on my laptop sends a message to Hong Kong, retrieves articles filed in Brussels, shows me pictures of my nieces in New York, and plays the song of a wood stork recorded in Florida. Traveling with my laptop and cell phone, I have access to a whole world of information and personal communication—a world that exists with increasingly little regard to geography, as electricity grids, phone towers, and wireless networks proliferate. This universe of instant information, conversation, and entertainment is so powerful and absorbing—and its currency so physically ephemeral—that it's hard to remember that the technology that makes it possible has anything to do with the natural world.

But this digital wizardry relies on a complex array of materials: metals, elements, plastics, and chemical compounds. Each tidy piece of equipment has a story that begins in mines, refineries, factories, rivers, and aquifers and ends on pallets, in dumpsters, and in landfills all around the world.

Over the past two decades or more, rapid technological advances have doubled the computing capacity of semiconductor chips almost every eighteen months, bringing us faster computers, smaller cell phones, more efficient machinery and appliances, and an increasing demand for new products. Yet this rushing stream of amazing electronics leaves in its wake environmental degradation and a large volume of hazardous waste—waste created in the collection of the raw materials that go into these products, by the manufacturing process, and by the disposal of these products at the end of their remarkably short lives.

Thanks to our appetite for gadgets, convenience, and innovation—and the current system of world commerce that makes them relatively affordable—Americans, who number about 290 million, own over two billion pieces of high-tech consumer electronics: computers, cell phones, televisions, printers, fax machines, microwaves, personal data devices, and entertainment systems among them.[9] Americans own over 200 million computers, well over 200 million televisions, and over 150 million cell phones.[10] With some five to seven million tons of this stuff becoming obsolete each year,[11] high-tech electronics are now the fastest growing part of the municipal waste stream, both in the United States and in Europe.[12] In Europe, where discarded electronics create about six million tons of solid waste each year, the volume of e-waste—as this trash has come to be called—is growing three times faster than the rest of the European Union's municipal solid waste combined.[13]

Domestic e-waste (as opposed to e-waste imported for processing and recycling) is accumulating rapidly virtually everywhere in the world that PCs and cell phones are used, especially in populous countries with active high-tech industries like China—which discards about four million PCs a year[14]—and India. The United Nations Environment Programme estimates that the world generates some twenty to fifty million metric tons of e-waste each year.[15]

The *Wall Street Journal*, not known for making rash statements about environmental protection, has called e-waste "the world's fastest growing and potentially most dangerous waste problem."[16] Yet for the most part we have been so bedazzled by high tech, adopted its products with such alacrity, been so busy thriving on its success and figuring out how to use the new PC, PDA, TV, DVD player, or cell phone, that until recently we haven't given this waste—or the environmental impacts of manufacturing such products—much thought.

Compared to waste from other manufactured products, particularly the kind we are used to recycling (cans, bottles, paper), high-tech electronics—essentially any appliance containing semiconductors and circuit boards—are a particularly complex kind of trash. Soda cans, bottles, and newspapers are made of one or few materials. High-tech electronics contain dozens of materials—all tightly packed—many of which are harmful to the environment and human health when discarded improperly. For the most part these substances do not pose health hazards while the equipment is intact. But when electronics are physically damaged, dismantled, or improperly disposed of, their toxics emerge.

The cathode ray tubes (CRTs) in computer and television monitors contain lead—which is poisonous to the nervous system—as do circuit boards. Mercury—like lead—a neurotoxin, is used in flat-panel display screens. Some batteries and circuit boards contain cadmium, known to be a carcinogen. Electronics contain a virtual alphabet soup of different plastics, among them polystyrene (HIPS), acrylonitrile butadiene styrene (ABS), and polyvinyl chloride (PVC). A typical desktop computer uses about fourteen pounds of plastic, most of which is never recycled. PVC, which insulates wires and is used in other electronic parts and in packing materials, poses a particular waste hazard because when burned it generates dioxins and furans—both persistent organic pollutants.[17] Brominated flame retardants, some of which disrupt thyroid hormone function and act as neurotoxins in animals, are used in plastics that house electronics and in circuit boards. Copper, antimony, beryllium, barium, zinc, chromium, silver, nickel, and chlorinated and phosphorus-based compounds, as well as polychlorinated biphenyls (PCBs), nonyphenols, and phthalates, are some of the other hazardous and toxic substances used in high-tech electronics. A 2001 EPA report estimated that discarded electronics account for approximately 70 percent of the heavy metals and 40 percent of the lead now found in U.S. landfills.[18]

In many places, solvents that have been used in semiconductor manufacture—trichloroethylene, ammonia, methanol, and glycol ethers among them—all of which adversely affect human health and the environment, have ended up in local rivers, streams, and aquifers, often in great volume. Semiconductor production also involves volatile organic compounds and other hazardous chemicals—including methylene chloride, Freon, and various perfluorocarbons—that contribute to air pollution and can potentially adversely affect the health of those who work with them. Numerous lawsuits have already been brought by high-tech workers who believe their health or their children's has been harmed by chemicals they were exposed to in high-tech fabrication plants.

Manufacturing processes and materials change continually and at a pace that far outstrips the rate at which we assess their environmental impacts—particularly in the realm of chemicals, where new compounds are introduced almost daily. Health and safety conditions throughout the high-tech industry have improved over the years, and the business has become more transparent. But the way in which the United States goes about assessing risks posed by chemicals used in high-tech manufacture has not changed, and many of the environmental and health problems now being dealt with were caused by events that took place over twenty years ago.

Despite the enormous quantity of electronic waste generated, and the fact that we have been producing this trash at accelerating rates since the 1970s, regulations and systems for dealing with this refuse have only recently been developed and put to work. In this, government policies regulating e-waste in the United States lag conspicuously behind those in Europe and Japan. As of this writing, about a dozen individual countries regulate the disposal of e-waste.[19] Over half of those have national systems to collect high-tech and other electronics products for recycling; the United States is not among them. As of 2006 it is mandatory throughout the European Union (although some countries have delayed compliance) and companion legislation restricts the use of certain hazardous substances in electronic products. A 2003 report by the International Association of Electronics Recyclers found that only 9 percent of Americans' discarded consumer electronics were being recycled.[20] Given the volume of electronics purchased and discarded in the United States, that we rely on voluntary measures to keep high-tech trash from harming the environment is like using a child's umbrella to stay dry during a monsoon.

And despite international regulations designed to prevent the export of hazardous waste from richer to less well-off countries, an estimated 80 percent of a given year's electronic waste makes its way from countries like the United States and the United Kingdom to poorer countries—like China, Pakistan, India, and those in west Africa—where huge amounts of equipment are dismantled in unsafe conditions or are discarded in ways acutely harmful to the environment.[21] No auditable figures are available, but industry experts estimate that about half a million tons of electronics are recycled in the United States annually.[22] Because this is no more than a tenth of what is discarded, somewhere between two and four million tons of e-waste from the United States alone has likely been making its way overseas each year for low-tech recycling. A recent study of e-waste in southern China found that about 75 percent of the electronics being processed there came from the United States.[23]

Some forty years have passed since Rachel Carson caught the world's attention with *Silent Spring*. A number of the synthetic chemicals Carson wrote about are now banned, but we continue to create new compounds with persistent adverse environmental and health impacts. Some of these manufactured substances are used to produce high-tech electronics, products that have become virtually ubiquitous throughout the developed world. Many high-tech electronics contain substances whose environmental impacts—local, global, short and long term—have not been dealt with before and which we do not yet understand. As we become increasingly dependent on the rapid electronic transfer of information, while telling ourselves that we are moving beyond the point where economies depend on the obvious wholesale exploitation of natural resources, we are also creating a new world of toxic pollution that may prove far more difficult to clean up than any we have known before.

That we have ignored the material costs of high tech is not surprising. Historically, industrial society has externalized many of the costs associated with its waste, expecting these costs to be borne not by manufacturers or purchasers of the products, but by communities and absorbed by the environment. Until the passage of clean air and water laws, industry could dump its effluent without expecting to be responsible for the consequences. In many ways high tech is a manufacturing industry like any other. But its public profile is very different from that of traditional industries. Because high tech enables us to store encyclopedias' worth of information on something smaller than a donut, we have—until very recently—overlooked the fact that miniaturization is not dematerialization.

As an illustration consider this passage from *Being Digital* by Nicholas Negroponte, published in 1995, which aptly characterizes the bloom and boom of high-tech culture and how our thinking about high tech tends to divorce the machinery from the information it transmits. (Caveat: In computer-chip generations, a statement about the Digital Age written ten years ago is like looking back at a view of the world penned during the

Roaring Twenties—and my aim is not to quarrel with or single out Mr. Negroponte for having made this observation.)

> The slow human handling of most information in the form of books, magazines, newspapers and videocassettes, is about to become the instantaneous and inexpensive transfer of electronic data that move at the speed of light . . . Thomas Jefferson advanced the concept of libraries and the right to check out a book free of charge. But this great forefather never considered the likelihood that 20 million people might access a digital library electronically and withdraw its contents at no cost.[24]

The point Negroponte, a professor of media technology at MIT, wished to emphasize was digital technology's potential to make information universally accessible, presumably without a cash transaction or equivalent thereof. Yet the phrase "at no cost" leaps out because it reinforces the perception that these digital gadgets perform their marvels with no material impacts whatsoever.

Where the garbage goes; where a plume of smoke travels; where waste flows and settles when it gets washed downstream; how human communities, wildlife, and the landscape respond to waste. These are costs traditionally outside the scope of the industrial balance sheet and that industry is just beginning to figure into the cost of doing business. As Jim Puckett, director of Basel Action Network, a Seattle-based nonprofit that tracks the global travels of hazardous waste, told me in 2004, "Humans have this funny idea that when you get rid of something, it's gone." The high-tech industry is no exception.

Laws regulating industrial waste have begun to protect human health and the environment from what comes out of chimneys and drainpipes, yet with few exceptions (e.g., state bottle bills) there is little mandatory collection of used consumer products in the United States. Manufacturers bear little responsibility for the post-consumer disposal of their finished products, and there are few industry-specific, legally binding bans on the use of toxic materials. But in the European Union, laws that become effective in 2005 and 2006 will require manufacturers to take back used electronics for recycling and to eliminate certain hazardous substances from their products. These regulations—known as the WEEE (Waste Electrical and Electronics Equipment) and RoHS (Restriction on the use of Certain Hazardous Substances) directives—are influencing what happens in the United States. Given the global nature of the high-tech industry, these materials standards will, in effect, become world standards, as it's simply not practical to have different manufacturing streams for individual markets.

High tech may thus become one of the first industries being seriously pushed to internalize the costs of waste throughout the products' life cycle and to design products with fewer adverse environmental impacts.

As of the end of 2005, the United States remains far from enacting any national e-waste legislation.[25] Yet over the past several years, more than half of all states have introduced some sort of e-waste bill. Meanwhile, most major high-tech manufacturers have set up some kind of take-back programs to facilitate recycling and reuse of their products. Manufacturers have also been teaming up with retailers, nonprofits, and local governments to hold used-electronics collection events. However, the burden of finding and using these programs still lies entirely with the consumer, and many are far more cumbersome, limited, and costly than comparable programs in Europe and Japan. And despite the fact that the United States has the highest per capita concentration of PCs, research published in early 2005 discovered that 95 percent of American consumers do not know the meaning of "e-waste" and 58 percent are not aware of an electronics recycling program in their community.[26]

However we cope with high-tech trash from now on, it's important to remember that many generations of this waste have already entered the global environment. As long ago as 1964 President Lyndon B. Johnson cautioned, "The bright success of science also has a darker side." We must, he said, "control the waste products of technology."[27] But virtually none of the books chronicling the rise of high technology or high tech's social and cultural influences consider the industry's impacts on human health or the environment. While knowledge of these impacts has existed for much longer, it has only been since the late 1990s that the world has begun to confront the environmental realities of high-tech manufacturing and e-waste in any substantive way.

Spurred by shocking pictures of this waste, the persistence of contaminated groundwater, serious health concerns about chemical exposure, and troubling scientific discoveries, we're scrambling to catch up. There are many reasons why we have allowed high-tech trash to pile up and pollute. Some are commercial: the historic practice of letting consumers and communities bear the burdens of waste. Some are political: the sway business and industry hold over public policy, particularly in the United States. And some are cultural: our embrace of the new, which seems to go hand in hand with our acceptance of all things disposable. It hasn't helped us come to grips with high tech's waste that when thinking about high tech many of us blur the distinction between hardware and software, forgetting that in addition to armies of computer-science jocks encoding the next operating system or search engine, high tech also means tons of chemicals, metals, and plastics. The problems created by high-tech trash, however, cannot be blamed on ignorance of the harm caused by industrial and chemical pollution, for by the time the high-tech industry came of age, professional knowledge and public consciousness of industrial pollution had been thoroughly raised.

The tangible effects of e-waste and the environmental and health impacts of high-tech manufacture may be out of sight for many people, but this is by no means a story of abstractions or problems so remote that they can be safely shelved. Nor is it a story that hinges on hair-splitting analyses of risk or an issue frothed up by worried advocates who yearn for simpler times. This is a story in which we all play a part, whether we know it or not. Information-age technology has linked the world as never before, but its debris and detritus span the earth as well. From product manufacture and marketing, raw material collection, order fulfillment, disposal and recycling—and because the cultures and politics of Europe, Asia, and the Americas influence what we consider waste and how we treat it, and because ecosystems do not respect political boundaries—this is an international story. If you sit at a desk in an office, talk to friends on your cell phone, watch television, listen to music on headphones, eat cheese bought in a supermarket almost anywhere in America, are a child in Guiyu, or a native of the Arctic, you are part of this story.

‖‖‖‖‖

This is probably a good place to interject that I am not a Luddite and that this book will not be an exercise in technology bashing. I am not anti-computer, I do not hate cell phones, abhor e-mail, or despise the Internet. Like most first-world citizens of the twenty-first century, I rely on these devices for much of my work, some entertainment, and personal communication. But I do not believe that "smart machines" and high-tech electronics can solve problems on their own or that they can replace human or natural creation and interaction. They are simply tools, to be used wisely and with inspiration, or not, as the case may be. The point of this book's investigations is not to condemn high technology, computers, and all their electronic relations, but to explore how the material demands of the digital age—as currently configured—are affecting the natural world and the health of human communities and how these problems are being addressed.

My interest in high-tech waste began a few blocks from my house, on the banks of the Willamette River in Portland, Oregon. In 2000 I wrote a report for Willamette Riverkeeper, a nonprofit river conservation group, investigating the toxics released directly into the Willamette. Thanks to decades of public outcry about the state of the river, many of the older pollutants—sewage, wood-products, and canning waste—that fouled the river for generations had been greatly reduced or eliminated. We wanted to find out how that progress might be holding up. Using information available through the EPA's Toxics Release Inventory (reporting required of industries that use over a certain volume of toxics monitored by the EPA), we discovered that between 1995 and 1997 alone the volume of toxics released directly into the Willamette Basin doubled, as did the amount of these toxics diverted to public treatment plants. The largest volumes of these chemicals came from factories producing semiconductors and from those processing metals, chemicals, and other materials for high-tech products.[28]

As in other communities around the country, high tech had been encouraged to settle in Oregon's Willamette Valley—part of the Pacific Northwest sometimes referred to as the Silicon Forest. The discovery that the high tech—a great economic, and in many ways, social and cultural boon to the Northwest and an industry commonly considered a "clean" or "green" alternative to timber and paper—was a major source of toxic pollution surprised all who read the report.

Since 2000, the world's awareness of e-waste and its impacts has burgeoned. Electronics recycling has become law in Europe and Japan, and manufacturers are racing to meet the European Union's 2006 deadline for eliminating certain toxics from their products. In 2002 ten U.S. states considered legislation concerning disposal of e-waste. In 2003 over fifty such bills were introduced. A handful of states have passed substantive legislation—among them California, Maine, Maryland, and Massachusetts. New studies on the impacts of chemicals used in high-tech products, on improvements in equipment design, manufacturing, disposal, and recycling appear almost daily. These topics have been the subject of intense debate on all sides of the Atlantic and the Pacific. Yet an enormous gap remains between what professionals and general high-tech consumers know about the hazards posed by e-waste and the environmental impacts of high-tech manufacturing, let alone the importance of solving these problems. I hope this book will help narrow this gap. For it seems to me that without this understanding we will continue to behave as if high-tech products exist in some kind of cyberuniverse, one that has little to do with the air we actually breathe, the water we drink, the food we eat, or our children's health.

The policies being formulated to deal with e-waste would not be under discussion in Brussels or Beijing or Washington, DC, if it were not for years of work, first by environmental and consumer advocates and then by legislators and business leaders who understand that the long-term sustainability of both industry and communities depends on making some important changes. As an environmental issue e-waste may lack the charisma of endangered species, ancient forests, and wilderness, but the push to put e-waste onto policy makers' agendas has similarly come from the grassroots—from concerned citizens and savvy NGOs in the United States, Europe, Asia, and elsewhere around the world.

The issue of e-waste has brought together activists from Hong Kong, India, London, California, the Philippines, the Netherlands, Texas, Wisconsin, and Seattle, with academics and researchers from China, Sweden, New York, and Tokyo, local and national government officials from around the world, and with executives from the world's leading high-tech manufacturers, retailers, and recycling and mining companies. There are tensions and great disagreements between environmental advocates and those representing industry and government, and between companies with very different corporate cultures, but this is by no means a story of good guys versus bad guys. It's more complicated than that.

I've been to half a dozen or more conferences on electronics recycling and related environmental issues since 2002, each attended by hundreds of industry professionals but by only a small number of environmental advocates. Without prodding from the environmental community, however, I don't think that any of the changes now taking place would be happening. And it's a charming irony that none of this global activity—on any side of the environmental activist or industry and government equation—would be possible or as effective without the aid of high technology itself.

In the conclusion of his book *Enough*, Bill McKibben lists what a number of influential thinkers consider to be the most significant innovations of the twentieth century. The two McKibben himself selects are nonviolence and wilderness. "Nonviolence, wilderness—these are the opposite of catalysts," he writes. "They're technologies that act as brakes, that retard our pell-mell rush forward, that set sharp boundaries on where we're going and how we'll get there. Right now, they aren't as important as computers. But one can at least envision a world in which they might be." And he continues, "We've been told that it's impossible—that some force like evolution drives us on to More and Faster and Bigger. 'You can't stop progress.' But that's not true. We could choose to mature. That could be the new trick we share with each other, a trick as revolutionary as fire. Or even the computer."[29]

Engineering and technology alone will not pull us out of the morass of high-tech trash. They must be accompanied by a desire to curtail the use of hazardous chemicals, stop the e-waste from piling up and from infiltrating the landscape, the atmosphere, the world's wildlife, and our bodies.

If we change our culture of instant obsolescence, our penchant for "More and Faster and Bigger," and our habit of ignoring the health and environmental impacts of manufacture until they have taken their toll, or our habit of tossing trash over the backyard fence in the high-tech arena, there will still be commerce, intellectual and scientific advancement, entertainment, electronic love letters, Listservs, digital relay of pictures, and wireless calls made to check on far-flung friends and family, but it won't be business as usual. Some changes in manufacturing, design, and disposal that will reduce the environmental impact of high-tech electronics are already under way. But a great many more need to be made. I set out to write this book with the hope of illuminating why such changes are so important—and because I believe that the more we know about the environmental and health problems caused by high-tech trash and high-tech manufacturing, and the wider this knowledge is spread, the more quickly these problems may be solved.

CHAPTER TWO

Raw Materials

Where Bits, Bytes, and the Earth's Crust Coincide

It's mid-July and the summer monsoons have begun, adding humidity to the 100-degree heat. Nobody in their right mind would be out in this mid-morning sun but I'm pulling into the Community Sports Center in Bagdad, Arizona, to meet my tour guides from the Phelps Dodge Mining Company. The two-and-a-half-hour-drive northwest from Phoenix has taken me through miles of stout saguaros and spiny Joshua trees. After the subdivisions petered out past Sun City, the Sonoran Desert reasserted itself and patches of irrigated lawn gave way to sand, cactus, mesquite, and creosote.

I've come to see the Bagdad Mine—one of the half dozen or more enormous open-pit copper mines scattered across the southern half of Arizona—because copper is an important ingredient of high-tech electronics. Arizona produces more copper than any other state—about 65 percent of the copper mined in the United States, or about 10 percent of the world's copper production. I want to get a sense of where the physical landscape and the world of bits and bytes coincide and a picture of where some of the stuff embedded in our PCs, cell phones, laptops, Palm Pilots, and other such gadgets actually comes from. It's a kind of time travel: tracing the materials of twenty-first-century technology back to their origins in the Precambrian and Cretaceous layers of the earth.

Of the slightly more than half of the materials in a typical desktop computer that are metals, the most likely to be found are copper, aluminum, lead, gold, zinc, nickel, tin, silver, and iron, along with platinum, palladium, mercury, cobalt, antimony, arsenic, barium, beryllium, cadmium, chromium, selenium, and gallium. Some metals—aluminum and iron, for example—are used structurally. Others, particularly the heavy metals (cadmium, lead, mercury, and other metallic elements that have high molecular weights), are used in circuit boards, semiconductors, and batteries.

Most heavy metals are toxic in low concentrations and tend to accumulate in the food web. Heavy metals cause neurological damage, adversely affect fetal development and reproductive systems, are known to cause kidney disease, and some are recognized carcinogens. Lead—one of the most commonly used heavy metals—can remain in the human body for years, lodging in the bones and circulating through the bloodstream. This is one reason you don't want old circuit boards to end up in landfills: these elements can leach into water and soil and seep into the local watershed, where they can be ingested by insects, fish, and other aquatic creatures and then work their way into our diet.

Some circuit boards use beryllium elements as electrical connectors and to insulate microprocessors. If improperly handled during disposal or recycling, beryllium dust—known to cause severe lung disease—may be released. Electronics recyclers I visited told me that given the lack of materials labeling, their workers had to identify and separate beryllium elements by hand.

Lead is used in computer monitor glass, television screens, and other cathode ray tube (CRT) glass to protect against radiation. The glass in a typical computer monitor or television screen contains between 2 and 3 percent lead. The frit—the part that creates the seal between the screen and the glass funnel that sits behind such screens—is 70 to 80 percent lead, while the funnel itself is between 22 and 25 percent lead.[1] CRTs themselves also contain lead as well as barium oxide. One large monitor may contain as much as eight pounds of lead. Often combined with tin and silver, lead is also used as solder to anchor various circuit board components. In a toxicity test, twenty-three out of thirty-three cell phones tested exceeded U.S. safety standards for lead.[2] Laptops, VCRs, printers, and remote-control devices put through similar tests for lead also exceeded safety standards. While equipment is intact, the lead doesn't pose a hazard, but when electronics are discarded and dismantled under any but the most controlled circumstances (i.e., state-of-the-art recycling), their lead is released to the atmosphere. Recent research indicates that very small amounts of lead harm children's cognitive development and that lead may be twice as toxic to adults as existing government standards assume.

The lamps that light flat-panel display screens—those illuminated by liquid crystal displays (LCDs), on laptops, newer desktops, and thin televisions—as well as some cell phones, batteries, circuit boards, digital cameras, and other hand-held electronic devices contain mercury. Even in small amounts mercury is known to cause damage to the brain, nervous, and reproductive systems, to the lungs, kidneys, and other organs, and to harm a developing fetus. It's also toxic to aquatic life and can work its way through the food web after being deposited in water bodies. An estimated 22 percent of the mercury used worldwide each year goes into electrical and electronic equipment, which includes batteries, flat-panel display lamps, and switches.[3] The Electronics Industry Alliance (EIA) has opposed legislation that would ban the use of mercury in the lamps used in high-tech electronics, reasoning that banning such small amounts would have only a "negligible impact on reducing mercury releases to the environment."[4] Using mercury, EIA explains, makes electronic products more energy efficient; if mercury were not used, more energy from coal-fired power plants would be required, resulting in greater mercury emissions.

Like other heavy metals used in electronics, mercury doesn't pose a health hazard while equipment is in use, but it becomes hazardous when equipment is disposed of improperly and it poses health hazards to workers during the manufacturing process. The U.S. Environmental Protection Agency estimates that about 4 milligrams of mercury are used to make the fluorescent light for each LCD and that each such unit produced is responsible for releasing approximately the same amount of mercury into the environment.* Mercury,

---

* The U.S. Occupational Safety and Health Administration considers any more than 0.1 milligram of mercury per cubic meter unsafe, while the EPA sets safety levels at one part per million in seafood and two parts per billion in drinking water.

says the EPA, can remain "in the atmosphere for up to a year and travel thousands of miles, potentially resulting in general population exposures."[5] Given the quantity of computers produced and discarded each year, electronics are likely contributing to the mercury circulating throughout the world and possibly ending up in the tuna sandwiches you and your kids ate for lunch.

Cell phones, newer computers, and some other high-tech electronics' capacitors make use of tantalum, a relatively rare ore often refined from coltan. While not poisonous in the way that mercury, lead, and other metals are, pursuit of coltan has had disastrous social and environmental consequences for some of the places where it's mined.

The remainder of a computer is made up of plastics, silica, glass, quartz, and various other nonmetallic elements and compounds that are used to create semiconductors and that make images light up on display screens and monitors. The majority of plastics used in high-tech electronics originate with fossil fuels. Few of these plastics are biodegradable, and many contain chemicals added for flame resistance, as coloring agents, or to impart a particular kind of strength or texture. Some of these additives have been detected in household and other indoor dust, as well as in the blood of people who live and work with electronics in nonindustrial settings. When burned or exposed to ultraviolet light (i.e., sunshine), many plastics break down into compounds that are toxic to humans, wildlife, and the environment.

The process of turning silicon wafers into microchips involves dopants—chemicals that make silicon semiconductive. These often include phosphorous and boron, polymer-based resins, and etchants such as nitrogen trifluoride gas and liquid hydrofluoric acid. This process also involves solvents, which are likely to include those made of volatile organic compounds, as well as those that are water based.

While the categories of materials are fairly consistent, it's hard to get a precise list of ingredients involved in the entire process of manufacturing a semiconductor or circuit board. The specific recipes for these products are generally proprietary, and materials vary depending on the kind of microchip being produced and its fabrication process. What's more, materials change continually as new products are developed.

Curious to know how these different metals, compounds, and other materials functioned within a piece of high-tech equipment, I consulted several well-known books about the computer industry and how computers work. Not one index yielded a single entry for any of the individual substances that enable this technology—further evidence of the curious disconnect between perceptions of high tech and the physical world. Delve into the guts of a computer or any other high-tech device and you'll discover that while more and more people are employed in endeavors that involve processing information, and fewer make a living extracting natural resources or in manufacturing, there's really no such thing as a completely "post-industrial digital economy."[6]

Even though the computer I'm using right now contains a relatively small amount of metal—far less than a car, refrigerator, or a house, let alone a bridge, skyscraper, or a new supermarket—extracting those ores from the earth's crust and turning them into the workable forms of metal that go into wires, conductors, and connectors leaves a large environmental footprint. Add in the impacts of processing the dozens of other materials—chemicals and plastics head the list—that go into each piece of high-tech electronics and that footprint is magnified and becomes considerably more complex.

Assessing these impacts involves not only considering the ecological footprint of one piece or component of high-tech equipment at a time, but also considering the vast quantity of semiconductors and circuit board–bearing devices that are being cranked out—and their short life spans. Each generation of high-tech equipment may be more efficient than its predecessor—in terms of performance and manufacture—but we're producing and discarding more electronics than ever while reusing only a small fraction of their materials. This pattern of consumption means more mining, more fossil fuels extraction, and more refining, with all of the direct and secondary environmental and health impacts that come with these processes.

The Semiconductor Industry Association estimated that in 2003 it would manufacture about ninety million transistors—the building blocks of semiconductors—for each and every person on the planet and that by 2010 this number would reach one billion.[7] This astounding rate of production is unprecedented by any previous generation of industrial goods and, given the material composition and materials intensity, has ecological ramifications not previously encountered. To make all of this less abstract, it seemed to me essential when assessing the impacts of high-tech products to connect the circuitry that enables cyberspace with its actual geographic origins in the natural world. Which is why I'm wearing a hard hat and standing in the baking sun on the edge of what looks like a huge terraced amphitheater carved into an Arizona hillside.

CONSEQUENCES OF COPPER

With a sweeping gesture, Phil Blacet, senior environmental engineer with Phelps Dodge, points out the perimeter of the mustard yellow pit that yawns in front of us. It's a mile and one-third across and one-third of a mile deep. Some dark jagged hillsides rise behind the far side of the pit, beyond where Blacet shows me the location of Boulder Creek. From there, high mesas stretch northwest, extending from the Upper Burro Creek Wilderness toward the Santa Maria Mountains. The elevation here is just over four thousand feet. The sky is blindingly blue. Trucks with tires nearly twelve feet in diameter trundle down newly carved roads that coil around the ochre colored slopes of the pit. Because the pit is so huge, the trucks are far enough away that, as big as they are, from where we stand I can hear only a gentle rumble of their engines.

About seventy million years ago, Blacet explains, magma from a prehistoric volcano he calls "Volcano Bagdad," crystallized and eroded. The result was a great swath of granitic rock riddled with minerals—copper, lead, silver, and a little bit of gold—that reaches across southern Arizona from Nevada east into New Mexico. Prospectors first came to Bagdad in the early 1880s. "They were looking for gold, but found copper instead," Blacet tells me. There has been copper mining here ever since. As I'm taking in all this geology, Bob Delgado, who worked for Phelps Dodge for years and now conducts public tours of the Bagdad Mine and is our driver today, bends down and hands me a dull turquoise blue rock. It's oxidized copper. As an inveterate collector of rocks, I want to pocket it but resist the urge.

Because copper is considered to be the best nonprecious metal conductor of electricity,[8] it's used in semiconductors, circuit boards, CRTs, high-tech telecommunications equipment, and in wiring for these and other electronics—both high-tech and those of earlier generations. The bulky computers with big monitors that grace many office desks may contain anywhere from two and a half to over four pounds of copper.[9] To get an idea of the quantity of just one of the metals high-tech equipment uses, consider these numbers: There are over 660 million computers in use worldwide[10]—over 200 million in the United States alone.[11] The Computer Industry Almanac reported in February 2003 that worldwide cumulative PC sales had surpassed 1 billion. The industry expects that number to grow by 61 percent by 2008. At a very rough estimate of two pounds per unit, that amounts to a lot of copper—more than 1.3 billion pounds in extant computers alone.

To put the desktop computer's copper contents in a larger industrial perspective, electronics and electrical products account for about 25 percent of the copper consumed annually worldwide. In the United States, electronics account for 20 to 25 percent or more of the copper consumed[12]— the only sector that consumes more is building construction. In Asia electronics consume some 50 percent (China is now the world's largest consumer of copper[13]) and in Europe about 37 percent.[14]

The volume of copper destined for electronics produced in the United States reached a high point in 2000. It declined significantly between 2000 and 2003 but increased slightly in 2004, mirroring the fortunes of the high-tech industry. But as Pete Faur, director of corporate communications for Phelps Dodge, explains, part of that decline resulted from manufacturing moving offshore. Less copper is also being used, says Faur, because manufacturing has become more efficient and the final products—particularly high-tech electronics—are getting smaller. Yet over the long term, worldwide copper consumption has grown substantially since the 1960s, and since the 1970s it has almost doubled.[15]

Most of the world's copper is now mined in South America, where Chile and Peru are the largest producers. North America comes next (with most from the United States, though the nation now imports about a third of the copper it uses[16]), followed by Indonesia and Asia—all rankings that may vary from year to year depending on production trends. Copper is produced in dozens of countries (although very little copper smelting now takes place in the United States), and a great deal of copper travels around the globe—large quantities are transported from Chile to China, for example—in its journey from raw ore to finished product.

The international movement of copper ore and refined copper is well documented, but it's hard, if not impossible, to trace a finished copper product from a specific mine to the computer sitting on your office desk. Not only does copper go through many manufacturing steps and travel multiple links in the supply chain, but the nature of this commodity market is less than transparent (figures concerning copper are often not publicly available and are highly proprietary, particularly in the case of super-competitive high-tech manufacturers).

Copper ore may have originated in the Andes in Peru, been shipped out to a smelter in Japan from a port north of Lima, the refined copper turned into sheets and wire in Europe, and teeny bits of the metal inserted into circuit boards in a factory in the Philippines. Some of the copper in your computer monitor may have come from a smelter in Sweden, where raw ore was mingled with copper taken out of used electronics that were dismantled in Rhode Island.

In addition to being an excellent conductor of electricity, copper is extremely recyclable.[17] "One hundred percent recyclable," says Ken Geremia, communications manager for the Copper Development Association. "What we like to say is that metals can be reused eternally because they don't get destroyed," explains Caisa Sameulsson, a metallurgist at the Minerals and Metals Recycling Research Centre at Sweden's Luleå University of Technology and who is developing new ways to recycle materials recovered from used electronics.

"Secondary" or scrap copper makes up slightly more than a third of the copper used worldwide.[18] With the proper collection and processing systems in place, and with the requisite social and political interest to make it happen, the rate of copper recycling could be about 85 percent.[19] In the United States about one-third of the copper used is scrap,[20] but less than 10 percent of this scrap copper comes from postconsumer sources.[21] The rest of the scrap is "new," the odds and ends of various manufacturing processes.[22] While about 90 percent of a computer's copper can be recovered and used again, only about 10 percent of high-tech electronics are recycled.[23] This means that about 90 percent of the copper that goes into PCs and similar electronics is never used again and, therefore, that most of the copper used in electronics is newly mined.

Mining is, in every respect, a costly business—as is the processing of ore into metal. Extracting ore from underground deposits requires vast amounts of capital, energy, water, and human resources. "It takes a long time—generations of people and investors to develop a mine like this," says Blacet as we peer into the Bagdad pit where several colossal trucks— big enough to scoop up some fifty-seven cubic yards of earth and rock per shovel load†—scrape methodically at the terraced slope.

Mining and smelting exact a heavy toll on the environment. Mining accounts for an estimated 7 to 10 percent of the world's energy consumption.[24] Most of this energy comes from oil and coal and is used to power the huge machines used throughout the mining and ore refining process. (In comparison, recycling copper uses 15 to 20 percent the amount of energy required to mine new ore.[25]) In the United States, mining releases more toxics than any other industry.[26] Smelters are the vast furnaces where raw ore is heated—usually to a melting point—to separate desired metals from impurities, including the sulfur often found with copper ore. Some of the impurities removed in this process are toxics that are usually released to the air and often include sulfur dioxide, nitrogen oxides, and lead. One notable case of mining-related air pollution occurred at the copper, lead, and zinc plant in La Oroya, Peru, where air emissions have caused virtually all the community's children to suffer lead poisoning. In Canada, in 2002, metals processors comprised half of

---

† When I got home I wondered just how much space fifty-seven cubic yards takes up. I figured out that my office, which is eight feet by nine feet with a nine-foot ceiling, would contain about twenty-four cubic yards—less than half a truckload at the Bagdad Mine.

the ten industrial facilities found to release more carcinogens into the air than any others.[27]

Some toxic by-products of mining and metals processing—arsenic, mercury, lead, and cadmium—travel with runoff into surrounding streams and groundwater. Some are deposited as solid waste in the form of tailings—the material that's discarded after the valuable ores have been extracted. A typical mine produces enormous quantities of tailings, which are typically piled in lengthy berms that resemble large earthwork dams, some as high as a thirty-story building.[28]

Tailings from copper mines contain sulfites and often a number of other metals, including lead, arsenic, cadmium, and zinc. When exposed to air and water, sulfites create sulfuric acid, which is very corrosive and is acutely toxic to aquatic life. For example, a bluegill fish will die if it is exposed to but 24.5 parts per billion of sulfuric acid in the course of twenty-four hours. Airborne sulfuric acid contributes to acid rain, which is toxic to plants, and inhalation of sulfuric acid mists has been linked to cancer of the larynx.[29] If washed into the watershed with the residual copper and heavy metals, the sulfuric acid creates a chemical combination that is similarly toxic to wildlife and to people who use the affected water.

The historic copper mining in Butte, Montana, has contaminated 120 miles of the Clark Fork River, which has become the nation's largest Superfund site. This contamination has reached the aquifer used by the communities of Bonner and Milltown, located at the downstream terminus of the site; residents there have not been able to drink their tap water for many years. And between 2000 and 2003, dozens of migratory birds died after drinking water contaminated by tailings from the Morenci Mine in Arizona.[30] Rain and stream water leach heavy metals from mine tailings into local watersheds to such an extent that the EPA estimates that some 40 percent of all headwaters and western watersheds—where most U.S. mining takes place—are contaminated by hard-rock mine pollution.[31]

Internationally, that mines have caused troubles both ecological and social is "a complaint heard from Nigeria to Papua New Guinea," as a 2003 *Time* magazine article put it.[32] But not all mines have such problems—some, like the Bagdad Mine, have been engineered to be what are called "zero discharge sites," closed or circular systems where water is recycled and doesn't leave the mine site. Still, a legacy of contaminated water is pervasive almost everywhere mining has occurred. And there is no getting around the fact that open-pit mining creates enormous unsightly craters, displaces huge amounts of soil, and disrupts the natural topographic ecology.

Most metals currently come from open-pit mines, many much bigger than Bagdad. To get a sense of their visual impact and size, after visiting Bagdad I drove down to see ASARCO's Mission Mine, south of Tucson. Its pit stretches two miles from north to south and one and three-quarter miles from east to west. Even bigger than the Mission Mine is the Morenci Mine—a Phelps Dodge mine that produces more copper than any other in North America. The mine, carved into the base of steep mountains north of Safford, Arizona, covers over three thousand acres, forming a gaping crater visible from miles away. About two-thirds of the world's raw metal—including nearly all copper and gold—comes from open-pit mines like these, which exist on practically every continent.

Excavation on such a gargantuan scale creates a correspondingly voluminous amount of waste rock and rubble. Producing one ton of copper from an open-pit mine like Bagdad or Morenci results, on average, in some 310 tons of waste rock and ore.[33] That would mean that going after the roughly 2 pounds of copper needed for a desktop computer would likely result in some 620 pounds of waste rock. By the same calculus, the two hundred million computers in use in the United States in 2005 have left in their wake 124 billion pounds of discarded rock—just to produce their copper contents alone. Substituting even 30 percent of this copper with new scrap or 10 percent with postconsumer scrap—the approximate rates for the amount of new and old scrap used in the United States—would still mean mountains of rubble. But the waste calculus would improve dramatically if the 85 percent rate of copper recycling the International Copper Study Group suggests is possible could be achieved.

"The problem is not the amount of postconsumer scrap available to work with, it's the logistics," says Theo Lehner of the Swedish mining company Boliden Mineral AB. Lehner is development manager at Boliden's Rönnskär smelter, which processes about thirty thousand tons of electronic scrap a year to extract copper and other metals. "If you don't make enough money, you leave [the scrap]," Lehner told me during my visit to Boliden. There is no problem in reusing a metal like copper, he explains. "With base metals, when they're processed you can't distinguish between those that come from a primary source [i.e., ore] and those that come from a secondary source—for example, that which has been recovered from a postconsumer or industrial source." And Lehner points out, "Production scrap is diminishing and postconsumer scrap is increasing as production moves overseas."[34] All of which strengthens arguments for not letting copper-laden high-tech electronics go to the dump, for collecting that postconsumer scrap, and for increasing the use of all kinds of recovered metals wherever possible.

THERE'S GOLD IN YOUR COMPUTER

In the fall of 2000, as part of a group of environmental journalists, I toured a gold mine in northern California, in the foothills of the Sierra Nevada. We looked at the mountainous berms of tailings, saw boxes marked "cyanide," peered at a pond filled with unnaturally turquoise water that in no way matched the late September sky, and looked down into the gaping cavity of the pit. We asked hard questions about environmental impacts of the mining and, although we were supposed to be unbiased, generally distanced ourselves from gold's symbolic lucre. Sensing the latter, a representative from the Gold Institute, a gold industry organization, asked us a question. "You all use computers, don't you?" he queried. "There's gold in computers," he told us with a note of triumph in his voice. This came as news to just about all of us.

There is indeed gold in computers, but don't pry open your computer hoping to find nuggets. An average desktop computer contains no more than an ounce of gold; a laptop, because of its size, somewhat less. This gold is dispersed throughout the circuit board in the form of incredibly fine gold wires that connect transistors, semiconductors, and other components. Circuit boards also have gold-plated connectors and contacts, as do plugs and sockets. Some integrated circuits use gold alloys as a bonding material, and some circuits are printed on a ceramic base using a paste that contains gold.

Gold is a good conductor of electricity and unlike copper or silver—the only better conductors—it does not corrode or tarnish, either of which can disrupt high-tech electronics' finely calibrated mechanisms. "Our age of high technology finds it indispensable," says the World Gold Council of their product. It's used "in everything from pocket calculators to computers, washing machines to televisions and missiles to spacecraft."[35]

Electronics products are currently the major industrial consumers of gold, accounting on average for about 10 percent of the world's annual gold production. In 2001 this amounted to about two hundred metric tons‡ and, as of this writing, about 7 percent of the gold used in the United States. Most of the rest goes first to jewelry and then to monetary investment, clearly dwarfing the amount used in electronics. Continuing miniaturization of circuit boards and increasing materials efficiencies in manufacturing mean that as semiconductors and high-tech electronics evolve, less precious metal—primarily gold, but also platinum, palladium, and silver—is used per piece of equipment. However, the ever-increasing volume of high-tech equipment means that the overall use of precious metals used in electronics continues to rise.

Like copper, most gold comes from huge open-pit mines. South Africa, Australia, and the United States are the countries that exhume the most gold, followed by China, Russia, Peru, and Indonesia, with significant amounts coming from elsewhere in Latin America, Africa, and Asia. Gold is mined on every continent except for Antarctica, where mining is not allowed. The environmental impacts of these mines and their often dangerous working conditions—particularly those of mines outside the United States—are well documented.[36]

Many gold mines, like the one I visited in California, use a process called cyanide leaching in which cyanide is sprayed onto raw ore to isolate the gold. This process makes it possible to extract gold from what is considered low-grade ore, the kind of ore now mined throughout much of

---

‡ A metric ton is 2,204.6 pounds, or 1.102 U.S. tons.

the American West. Liners are placed under enormous piles of ore, but leaks and failures have occurred, causing severe toxic contamination of surrounding soil, streams, and groundwater and consequently of everything in the local food web.

In many places, both in the United States and abroad, the dumping of mine waste has rendered surface and groundwater undrinkable and has turned water acidic and lethal to fish and other wildlife. Some mine waste has been dumped in coastal waters, contaminating the marine environment. In addition to causing water pollution, open-pit mines can cause great shifts in local water supplies, often dewatering adjacent streams and sometimes even the immediate aquifer. This can happen if the depth of a mining pit is lower than the water table. The pit must then be pumped dry to keep it workable, while large amounts of water from the same source are used concurrently to control dust and for other mining operations.

The environmentally correct may eschew gold jewelry, but when it comes to high-tech devices, choosing to go gold-free (at least for now) is probably not an option. However, with the proper systems in place for recovering and processing used electronics, the gold that goes into high-tech equipment could easily come entirely from recycled, previously mined and used gold.

Gold, like copper, is in theory 100 percent recyclable. "Gold is virtually indestructible," says the World Gold Council, "so that, unless it has been lost, all the gold ever mined still exists."[37] Most of this gold is in fact still with us—the majority of it in the form of gold bars, coins, and jewelry. Reflecting on the traditional stockpiling of gold, the economist Robert Triffin remarked, "Nobody could ever have conceived of a more absurd waste of human resources than to dig gold in the distant corners of the earth for the sole purpose of transporting it and reburying it in other deep holes."[38] I wonder what Triffin would say about tossing gold-bearing high-tech electronics into the trash.

The World Gold Council estimates that in 2001 about two hundred metric tons of gold went into electronics and electronic components.[39] Assuming that about two-thirds (about 294,000 pounds) of this gold was newly mined—using the U.S. Geological Survey (USGS) estimate that one unit of gold mined results in 2.7 units of waste—this amount of gold would have produced almost 794,000 pounds of waste. When properly dismantled and processed for recycling, however, it's possible to recover about 99 percent of the gold in a typical desktop computer—just about all of which is located somewhere in the circuit board. And it was the allure—and value—of precious metals recovery that provided some of the initial economic incentives for recycling old pieces of high-tech equipment and that prompted a number of major mining companies to get involved in electronics recycling.

"One metric ton of circuit boards can contain . . . 40 to 800 times the concentration of gold contained in gold ore mined in the United States," says the USGS.[40] The same agency estimates that one metric ton of discarded PCs contains more gold than can be recovered from seventeen metric tons of gold ore, making discarded electronics a more reliable source of gold than mining. Yet even with gold's recyclability and its high value, only about 30 percent or so of the gold used throughout the world comes from scrap. And the majority of gold that gets recycled is scrap from jewelry rather than from used electronics. Imagine if it were possible to retrieve every ounce of gold from all of the world's 660 million computers. That would yield about 41.25 million pounds of gold, which from a mining company's point of view is infinitely more accessible than ore that must be blasted out of a mountain. Instead, most of that reusable precious metal is ending up in landfills, and open-pit mines continue to be worked all over the world.

KILOMETERS UNDERGROUND:
ZINC, SILVER, AND OTHER METALS

Some gold and copper—as well as silver, zinc, lead, nickel, and other metals— comes not from open-pit but from underground mines. One such site is the Renstrom Mine operated by the Swedish mining company Boliden, which is the world's fourth largest producer of zinc and which runs the third largest copper smelter in Europe. I've come to Sweden to visit the Boliden operations because the company is pioneering new ways of processing the copper extracted from discarded high-tech electronics.

The mine, which has been in operation since 1952, is located about forty kilometers from the northern Swedish coastal city of Skelleftea. To reach the mine from the city I drive west and inland, upstream along the Skelleftea River from its mouth at the Baltic Sea port of Skelleftehamn. If one were to invent a landscape completely opposite to Arizona's copper mining districts, this would be it. In late May it's damp and lush with early spring undergrowth. The rivers—some of which rise north of the Arctic Circle, only a hundred kilometers or so north of here—are running full with snowmelt. On my way to the mine I pass green farm fields, wind whipped lakes, woodlands of pine and birch, and a lot of small logging operations.

At this time of year there are about twenty hours of daylight at this latitude, but the deciduous trees are just beginning to leaf out. The weather is chilly and rainy, and there has been an almost ceaseless blustery wind. Under these low overcast skies, the light is the same pale gray almost twenty-four hours a day. The water surfaces are a steely blue, the barely visible birch leaves apple green, and the pines a deeper fir color. The houses and barns are mostly painted red with white trim. A few are a rich bronze mustard color. Before I reach the mine site itself, I pass another road that leads to a tailings pool banked high with gravel and mostly hidden by surrounding woods.

From aboveground the mine looks only like a small quarry of sorts, with work buildings, some heavy machinery, and some gravel piles. Near the main building I'm greeted by a large man in blue coveralls who is wearing work boots and a hard hat with a miner's lamp. Strapped to his person are all sorts of radio gear, including walkie-talkie and cell phone. He tells me his name is Kurt and that he's an environmental manager at the mine. "So you want to see the mine," he says. He shows me to the women's lockerroom, where I find a pair of work boots that fit, a padded blue coverall, and hard hat.

We proceed to the building that houses the mineshaft, where we board an open, unlit elevator. As we go hurtling down—it's like being in a highspeed dumbwaiter—there's a damp clay smell and the sound of water dripping. It's cold and very, very dark. We rattle down at what seems like a perilous speed to what turns out to be a depth of 808 meters. "I climb the shaft twice a year to make sure the steps are all right," Kurt tells me as cold, clammy air whooshes through the shaft. We step out and I can see the entrances to what appear to be a warren of tunnels. The walls around us are reinforced by concrete and iron, but dripping with water. The muddy clay smell is even more intense.

In a moment we're met by a woman named Katrin in a pickup truck who hands me a pair of squashy plastic earplugs and steers expertly through the underground maze. The tunnels are lit only by intermittent bulbs, and the walls are marked with spray-painted numbers and arrows. We drive farther down into the earth, through forks in the maze of tunnels, to a depth of 1,038 meters. We stop where a man has parked a large yellow piece of equipment that has "Rocket Boomer" printed on its side. Katrin drives off, and all of a sudden it dawns on me that we're more than a kilometer underground.

The man operating the Rocket Boomer machine is drilling small holes in the face of the wall where explosives will be placed to blast out the rock to enlarge the work area. The cavern where we're standing must be over twenty feet high. Zinc, silver, gold, and perhaps some lead and copper are what the rocks here contain. Approximately 15 percent of the world's silver is now used by the electronics industry.[41] This percentage may soon increase because solder made of silver and tin is being considered as a substitute for the lead solder traditionally used in circuit boards.

As we're waiting for Katrin to return and take us to another part of the mine, Kurt points out what looks to be a large green box. It's a "refuge" he tells me. "The most dangerous thing down here," he says, "is fire." If there's a fire, he continues, "We can go into the refuge where there is fresh oxygen for up to eight hours." Somehow I don't find this entirely reassuring. I distract myself from a moment of incipient claustrophobia by reminding myself that, every day, dozens of people work deep underground and that I'm here because bits of what are being blasted out of these underground rock faces may be in the computer with which I'm going to type my notes.

Our next stop in this subterranean universe is an alcove where two men are drilling samples to be tested for their ore contents. We walk in through puddles, stepping over pieces of equipment, cables, and pieces of rebar. The reinforced rock and mud walls are running with water. The men are working near a set-up of tools, on top of which sit wooden boxes with narrow slot shelves in them. They look like the boxes that old move-able type was stored in. These boxes hold the drilled samples—cylinders of rock extracted from the wall. Kurt explains that this kind of work goes on continuously here. Small samples are analyzed, then, if they prove suf-ficiently ore laden, the big machinery follows to blast out large quantities. "Drill and map" is the process, Kurt tells me.

Later, back aboveground, Kurt shows me video taken by computer-operated cameras that monitor the conveyor loads of rock and ore com-ing out of the mine. The zinc will be processed at Boliden's smelters in Finland and Norway. Ore containing copper, precious metals, and lead goes directly to Boliden's Rönnskär smelter in Skelleftehamn. There it will be mixed with metals extracted from shredded circuit boards that come from used electronics collected not only in Sweden and the rest of Scandinavia, but also in other parts of Europe, North America, South Africa, and Malaysia. The geography is a bit mind-boggling: a computer discarded by an office in California or Boston may end up in a cauldron with ore mined under the boreal forests of Sweden.

Thanks in part to all the high-tech automation, and to careful practices, Boliden's Renstrom Mine is safer and more environmentally sound than many of the world's other underground mines. Aside from the fires, tun-nel collapses, and soil destabilization most often reported on, the greatest health risks of underground mines, says the United Nations Environment Programme (UNEP), "arise from dust, which may lead to respiratory problems."[42] And while these mines are largely hidden from view, they use large amounts of water and energy, and their operations create tailings, which like those of surface mines get piled in berms. Once again we're left with a good argument for increased reuse of metals, as recycling uses only a fraction of the resources required to make the same item from newly mined ore and creates far fewer health hazards.

FROM SAND TO SEMICONDUCTORS

When one thinks about high tech, it's typically not lead, silver, gold, or copper but silicon that comes to mind. The words "silicon" and even "Silicon Valley" have become stand-ins—what poets and linguists might call a metonymy or synecdoche—for the entire high-tech industry. And while some other materials are used to make semiconductors, silicon wafers do indeed form the basis for nearly all the transistors and semi-conductors that drive high-tech electronics.

Silicon, a kind of sand, is a common and simple substance, but turning it into the hyperpure polished bits of wafer that carry digital impulses—the heartbeat and brain waves of the Information Age—is a complex process that involves many toxic chemicals and that creates numerous waste products. On its own, silicon is not electrically charged, but its chemical structure makes it ideally suited to transformation into a semiconductor—a device that can be made to carry highly sophisticated patterns of charges by adding various chemical "impurities." This property is at the heart of what enables the computer on which I'm typing to put words on the screen or find a document in its memory. Getting a glimpse of the complexity of wafer manufacture seemed to me key to understanding the materials- and energy-intensive nature of high-tech electronics.

Given the enormous number of semiconductors and their growing ubiquity, one might think that the high-tech industry gobbles up most the world's supply of silicon. This, however, is not so. The semiconductor industry uses only about 2 to 5 percent of all the silicon used industrially.[43] The rest of the world's silicon gets used by the steel and ferrous metals industries, by aluminum producers, and by the chemical industry, a portion of which in turn fuels part of the high-tech electronics supply chain. So, while it would be very difficult to measure, more silicon goes into high-tech electronics than is used in semiconductor wafers alone. And some of the hyperpure sand that goes into the silicon wafer production process comes from what is left behind by the steel industry, so silicon wafers contain both new and recycled (although not postconsumer) silicon.

Work leading to the production of high-purity silicon became the basis of high-tech electronics began in the late 1940s and early '50s. "We started playing with silicon around 1953," says Jim Moreland, vice president of strategic development at Siltronic—one of the world's major producers of silicon wafers—speaking from his office in Portland, Oregon. Siltronic has plants around the world, in Germany, Singapore, Japan, and the United

States. Its North American plant, located in Portland, Oregon, makes wafers but no longer grows the silicon crystals, which the company makes in Germany and then transports to the United States for further processing.

Silicon wafers themselves were first made in the early 1960s. Siltronic, which sells wafers to IBM, Intel, Motorola, and others, puts that date at 1962, about four years after production of the first semiconductors. Some of the first wafers—like those made by Motorola—were square, Moreland explains, but there were problems with handling the squares, so the industry began to grow round wafers and has remained with that shape since the mid-1960s.

"The only thing in this industry that's been standardized is the diameter of wafers," says Moreland, explaining how the silicon wafer production process changes continually in response to demands of semiconductor manufacturers. Depending on the final application—what kind of semiconductors will be built on the wafers and what kind of equipment the chips are to run—there are different kinds, sizes, and categories of wafer.

"All the wafers are custom-built to meet specific parameters and requirements," says Moreland, with specific impurities added for conductivity. These added chemical impurities are what make silicon wafers—otherwise not conductors of electricity—semiconductors. The impurities, also known as "doping additives" or "dopants," include antimony, arsenic, boron, and phosphor—all highly toxic elements, particularly in the chemical forms used for wafer processing. "For special applications," says the Siltronic Web site, "neutrondoped crystals are produced in nuclear reactors, using radiation."[44]

The raw material that provides the silicon used to make semiconductors is actually silicon dioxide, a compound of silicon and oxygen. This silicon usually comes from high-purity quartz or quartz sand, which happens to be the second most abundant element in the earth's crust (oxygen being the first and aluminum the third). The industrially useful forms of silicon and silica are often quarried as ferrosilicon, which is mined all over the world. Argentina, Australia, India, Iran, and South Africa are just some of the countries that produce silicon; even Bhutan and Bosnia-Herzegovina are listed among the ferrosilicon-producing nations. According to the USGS, China is by far the world's largest producer of ferrosilicon, producing nearly twice as much as Russia or

Norway, the next two largest producers.[45] Altogether, millions of metric tons of ferrosilicon are produced worldwide each year.

Except for temporarily disturbing the immediate area while mining operations are active, the sand and gravel removal involved with silica mining usually has limited environmental impact."[46] On this point environmental advocates and the mining industry agree, pointing out that while silica mining disturbs ground (think enormous gravel pits), which can cause damage— sometimes considerable damage—its impacts are less than those of mining metals like copper or gold. Unlike hard-rock mining for precious metals, silica mining does not require a leaching or precipitation process involving applied acids, and therefore does not create tailings.

Silicon dust, however, is hazardous, and when inhaled—especially over long periods of time—can cause a lung disease known as silicosis for which there is no cure. Silica dust can cause scar tissue or fibrosis to form in the lungs, reducing their capacity to process oxygen. This leads to fatigue, shortness of breath, and susceptibility to infections and other lung diseases, like tuberculosis and emphysema, and eventually respiratory failure. Early stages of the disease are hard to detect, and chronic silicosis usually shows up ten or more years after a person's exposure to the dust.[47] According to the Occupational and Safety and Health Administration, active prevention programs in the United States have caused rates of silicosis to decline sharply over the past thirty to forty years, from about 1,160 deaths in 1968 to about 190 in 1999, with the most cases occurring among mine workers and those working in manufacturing industries.[48] But internationally, the World Health Organization reports that thousands of people die from silicosis each year—mostly those who have worked in mining and construction.[49] Crystalline silica dust occurs a long way from where silicon wafers are made, but it seems worth remembering the origins and consequences of the raw material that becomes a product renowned for its purity.

|||||| ||| ||

The silicon that becomes the wafers onto which semiconductors are etched is of exceptionally high quality.[50] In the first step of water production, silicon dioxide is ground up into a fine powder that is then distilled in a process using carbon dioxide and hydrogen chloride to further purify the silicon. What results is a colorless chemical called trichlorosilane, a highly toxic liquid—flammable and corrosive—that causes severe health problems if it comes into contact with skin or eyes or is inhaled or ingested. Its odor is described as pungent and suffocating.

This liquid is then heated until it reaches a vapor stage and the gas then reduced with hydrogen to remove the chlorine. What's left behind is silicon that is 99.999 percent pure, in a form called polysilicon, Moreland explains. This is what is used to grow silicon crystal rods, which look like big, supershiny silver or chrome rolling pins. Chunks of polysilicon that have been cleaned still further are put into special quartz crucibles and heated until they change from a solid to a liquid. A single seed of silicon crystal is placed into this liquid, positioned precisely to create a surface tension that directs the growth of the crystals into the desired shape.

As the ultrapure crystal silicon is grown, it is rotated and baked until the desired diameter is attained. The crystal is then cooled and ground into a perfectly cylindrical shape. The crystal is now ready to slice into wafers. Imagine a seriously high-tech, ultrapure silicon crystal roll of refrigerator cookie dough.

To achieve the phenomenal degree of smoothness and lack of variation from wafer to wafer required by the semiconductor fabrication process, the flat wafers are cleaned and polished after they are cut. Part of the process of polishing the newly cut wafers involves multiple steps of etching and washing that use both acids and base—or caustic—solutions to remove what's called the "damage" layer created by the mechanical slicing. As Myron Burr, environmental engineer at Siltronic, explained it to me, "The damage layer is removed by dissolving the top layer of silicon by a caustic etch process using potassium hydroxide followed by an acid etch process using nitric acid and hydrofluoric acid."[51]

After the polishing is finished, a laser is used to inspect wafers for defects—defects that may be as small as 0.12 microns, or perhaps even smaller.§ Some wafers then have an additional layer of silicon laid on as a foundation for

---

§ A micron is one-millionth of a meter. Ten microns are often described as being the size of one-seventh the diameter of a human hair.

highly sensitive transistor etching. The finished wafers are then shipped to a semiconductor manufacturer where the process of turning them into microprocessors takes place. A single wafer—the largest ones are now 300 millimeters (nearly a foot) in diameter—will produce dozens of chips.

Producing hyperpure silicon from raw silicon and turning it into finished wafers for semiconductor fabrication is a materials- (and energy-) intensive process that consumes a volume of raw material considerably larger than that of the finished product. According to one life-cycle analysis, 9.4 kilograms of raw silicon are used to produce 1 kilogram of finished wafers.[52] This would be as if you needed nearly nine and a half cups of water to make just one cup of tea.

The chemical contents of the wastewater from this step of etching and cleaning the silicon wafers are what first drew my attention to Siltronic's Portland, Oregon, production facility. In 2000 I wrote a report on the point-source pollution entering the Willamette River and its tributaries using the EPA's Toxics Release Inventory (TRI).** This data showed that in 1997—the most recent figures available in 2000—Siltronic (or Wacker Siltronic as the company was then called) had released nearly 1.3 million pounds of toxics, mostly nitrate compounds, directly into the Willamette River. This made Siltronic 1997's largest direct discharger of toxics to the Willamette. Over the next several years, Siltronic's discharges of toxics to the Williamette hovered around a million pounds annually—with a high in 2000 of slightly over 1.3 million pounds. In 2003 (the most recent data available as of this writing), Siltronic remained the largest direct discharger of nitrate compounds to the Willamette, having released some 750,000 pounds of these compounds.[53]

** TRI requires businesses to report annually the release of certain chemicals used by businesses at specified volumes. It is part of the Emergency Planning and Community Right-to-Know Act designed to help local communities protect public health, safety, and the environment from chemical hazards, which was enacted in 1986 in response to the 1984 release of chemicals from a Union Carbide plant in Bhopal that killed thousands of people and "a serious chemical release from a sister plant in West Virginia."[54]

When I asked about the large volume of nitrates discharged to the Willamette, the company's environmental engineer, Myron Burr, explained that nitric acid is one of the acids used in the wafer-etching process. "All of the process wastewater is treated to remove the fluorides and dissolved silicon," he said, "after which the acids and caustics are neutralized before they can be released as required" under the company's state and federal discharge permits.

"This neutralization process," Burr continued, "creates water-soluble nitrates. Nitrates are not removed in the treatment process because all forms of nitrate salts (neutralization products) are soluble and pass through any treatment system. Fortunately," he added, "nitrates are used by plants as nutrients." This is true. However, excess nitrogen deposited as a result of human activity—largely from sewage and agricultural runoff, but also from fossil fuels and industrial releases—also prompts excessive aquatic plant growth. This excess growth often occurs as algae blooms and can alter the nutrient composition of aquatic environments and upset the balance between native aquatic plants and animals. The effects of excess nitrogen have become a problem for ecosystems worldwide.

The company has "recognized the potential for excess nutrient loading and [has] spent several years working on process improvements, treatment options, and reuse options to minimize the amount of nitrates from wafer manufacturing," Burr said. As a result Siltronic has been working to reduce the acid used to etch each wafer and also to extend the life of each etching bath as well as the acids and caustics used in the baths. Siltronic has also been exploring changes and the possibility of making some products with an etching process that doesn't use acid.

As part of these efforts Siltronic developed a system in which another company acquires Siltronic's used nitric acid to use as a metal etchant. This enabled Siltronic to reduce its nitrate discharges by over 80,000 pounds in 2003. "In the first four months since implementation in 2004," Burr told me, "we have successfully reduced the nitrate discharges by approximately 145,000 pounds, or about 40 percent;" and in 2005 "we plan to reduce the nitrate discharges by 400,000 pounds." Reusing the acid, Burr added, also reduces the amount of chemicals required to treat the nitric acid, and thus the waste generated by this treatment, including "the greenhouse gases required to manufacture the displaced nitric acid in the first place."

Siltronic has been making wafers in Portland since about 1979, and their recent environmental improvements are impressive. But without detracting from those achievements, it's worth noting that Siltronic's operations have released into the Willamette close to twenty-five years' worth of substantial volumes of nitrate compounds—among other toxics. And Siltronic is far from the only high-tech manufacturing firm that has been sending toxic effluent into the Willamette Basin. Data accessible through the EPA's TRI database in 2000 showed that in addition to Siltronic, Mitsubishi Silicon America (silicon wafers), Hewlett-Packard, Intel, Praegitzer Industries (printed circuit boards), Merix Corporation (printed circuit boards), Integrated Device Technologies (semiconductors), and Wah Chang all released toxics into local surface water, either directly or by way of a treatment plant.

Many of the high-tech companies in the Willamette Valley and elsewhere that manufacture silicon wafers, circuit boards, or semiconductors also release toxics into the air. Some of the toxic air emissions resulting from silicon wafer fabrication include ammonia—a solvent used in one of the wafer-polishing steps—hydrochloric acid, hydrogen fluoride, and nitric acid. These releases, it should be noted, are all legal and permitted and are typical of the industry wherever it's located, but that doesn't make the process of rendering silicon hyperpure a correspondingly "clean" industry. The Willamette is also far from the only river system in the United States into which the high-tech industry releases its effluent, and Siltronic is but one of many companies located all around the world making silicon wafers.

That said, since the late 1990s Siltronic—albeit one company among many—has managed to reduce its use of water and chemicals considerably by introducing various recycling systems and modifying manufacturing processes. Since 1997 the company has reduced what it calls "its application-specific consumption" of hydrofluoric acid by 50 percent and of nitric acid and other cleaning materials by 70 percent.[55] While Moreland says Siltronic has not been able to find a use in the United States for the waste material generated during the wafer slicing process, it has been able to do so at its plant in Burghausen, Germany. There, the sludge or slurry gets broken down into its components; the oil is used as a fuel to make bricks and clay products, while the solid components—silicon and silicon carbide—are used in another industrial process. At another of its German plants and at its plant in Singapore, Siltronic has found a way to reuse an alkaline material employed in one of the final wafer-polishing steps to reduce its consumption of this material by 92 percent, the company reports. Siltronic has also developed a nontoxic solvent substitute for the solvent that had been the source of the large volumes of ammonia discharged up until a few years ago. While Siltronic—and other high-tech manufacturers—continue to make environmental improvements[56] and become more transparent in their reporting on resource use and emissions, given the complexity, scale, and global reach of these operations, gaps remain in our ability to assess the industry's ecological footprint.

▐█▎▌█▐ █

Among those who have been trying to produce a comprehensive assessment of the resources required to produce a silicon wafer destined for semiconductor fabrication are United Nations University researcher Eric Williams[57] and colleagues. By their calculations, one square centimeter of finished silicon wafer weighs about 0.16 grams—or about half the weight of a hummingbird egg, which is about the size of a small jelly bean (but considerably lighter). Williams's team estimates that making this small and incredibly delicate item requires about twenty liters of water. "A typical 6-inch wafer fabrication plant processing 40,000 wafers per month reportedly consumes 2–3 millions of gallons [of water] per day," they wrote in 2002.[58]

Williams's team calculated that producing one square centimeter of wafer also requires forty-five grams of chemicals (over 250 times the weight of the wafer section being produced) and 556 grams of elemental gases (nearly 3,500 times the weight of the wafer section). This processing also uses about 1.8 kilowatt hours of energy—enough energy to run a 100-watt lightbulb for eighteen hours. Approximately one-sixth of this energy, estimate Williams and his colleagues, comes in the form of direct fossil fuels, the rest in electricity. In addition to the energy required to turn the raw material of silicon dioxide into a polysilicon crystal and slice it into wafers, energy is needed to heat, ventilate, and provide air conditioning to clean rooms where parts of this process take place.

The silicon wafer fabrication process, also according to Williams and his colleagues' calculations, generates some 17 kilograms of wastewater and 7.8 grams of solid waste. To put these numbers in perspective, it's worth remembering that an entire 200-millimeter wafer is about 314 times one square centimeter. Other researchers' analysis I consulted used different units of measure but reached comparable conclusions.[59]

Williams and his colleagues note that the variety of data sources for this information and the proprietary nature of the industry pose a considerable challenge to those trying to produce an accurate life-cycle analysis of high-tech manufacturing. Still, the numbers themselves are actually extremely important because they are the measures manufacturers will have to work with in order to evaluate their progress in reducing natural-resource use and waste production. From my perspective, however, the point of these assessments is not so much the exact number crunching, but the understanding to which the analyses contribute: that manufacturing the high-tech electronics that bring us a seemingly immaterial world are in fact great consumers of the material world itself.

Wondering about the environmental impact of the finished wafer or a finished chip, I asked Moreland for his thoughts on the subject. From a purely physical size perspective, Moreland pointed out that "a chip is a small part" of a piece of high-tech equipment. "At the chip level, there isn't really a way to reuse them because chips are so specific to a task and because of the obsolescence of their capability," he said. "How to deal with the chips themselves? It's pretty benign," he told me. "Any doping is locked into the crystal matrix, so landfill is the best way to dispose of them. Silicon oxidizes over time, and even though it's not found in the environment in the state we use it, and I doubt that it's biodegradable, there are no long-term studies that show that it does any harm. You could swallow a chip and it would not hurt you," he added.

While it may be true that a chip could pass through a human without doing harm—I, personally, would not recommend trying it out, either by chewing a chip or by gulping it down like a pill. And once part of a circuit board's central processing unit (CPU), a semiconductor does not exist in isolation but it is attached to the circuit board with gold and copper wiring and often other precious metals. (This is why in the low-tech, primitive recycling of circuit boards the boards are melted down or smashed to get at the chips.) There is also likely to be lead solder involved, and there may be a beryllium element nearby, not to mention the plastics into which all of this is embedded.

The current design of high-tech electronics makes it highly likely that individual microchips and bits of silicon wafer embedded in circuit boards will indeed end up in landfills (or incinerators) unless these electronics are professionally recycled. (There's simply no other way to safely extract individual components for recycling.) And given our repeated experience with any number of the substances associated with high-tech electronics—trichloroethylene, polybrominated diphenyl ethers, and perfluorooctane sulfonate to name but a few—the lack of long-term studies showing whether silicon wafers or semiconductor chips pose environmental harm if deposited in landfills isn't too reassuring.

So the next time you wiggle your toes in the sand at the beach, think about how this soothing substance can be transformed into a material that enables the frenzy of e-mail, pop-up ads, garish high-definition television screens, PowerPoint presentations, and jangling cell phones that many people flee to the beach to avoid. Yet to be fair, the transformed silica also makes it possible to take the laptop out on the deck or to the coffee shop instead of working cooped up in an office.

Turning simple silica into the platform for nearly all high-tech electronics is anything but simple. It requires enormous amounts of other materials, highly complex machinery, energy, and water, and creates large amounts of waste. Like most of the high-tech manufacturing processes, the environmental impacts of silicon wafer fabrication have received little attention from the general public, but they are global. This manufacturing takes place in

China, Japan, South Korea, the Ukraine, Germany, France, Singapore, Taiwan, Turkey, India, Russia, Malaysia, the United Kingdom, and Indonesia, as well as all over the United States. So a description of the manufacturing process at a company like Siltronic has to be multiplied many times over, with the geography of impacts fanned out around the world.

TANTALUM PLUS COLUMBIUM EQUALS COLTAN

As ubiquitous as the primary raw material of silicon wafers is, a substance that goes into capacitors that are used in cell phones, laptops, and other high-tech electronics is little known. That substance is coltan, whose origins provide a strange cautionary tale about the global supply chain and the source of raw materials that go into high-tech products. Mined on several continents, this mineral can link a cell phone, video game console, or digital camera to a miner standing waist deep in muddy water in the northeastern reaches of the Democratic Republic of the Congo, to a refinery in Kazakhstan that receives ore shipments by way of the Middle East, and to a metals processing company headquartered in Pennsylvania. Coltan, wrote Blaine Harden in the *New York Times* after traveling to the eastern Congo in 2001, represents "a squalid encounter between the global high-tech economy and one of the world's most thoroughly ruined countries."[60]

Coltan is a combination of two ores, columbium (an element originally called niobium) and tantalum, which are found together in most of the rocks where they occur.[61] Coltan is the major source of tantalum, an extremely valuable and useful metal. Tantalum is highly heat and corrosion resistant and is an excellent conductor of electricity, qualities that are ideal in tiny capacitors—devices that store an electric charge much in the way a water tank stores water.[62]

Capacitors serve as an interim storage place that can be drawn upon when supplies dwindle at the source, hence a capacitor's usefulness in portable electronics. Tantalum capacitors have become a vital part of the digital circuitry found not only in cell phones and laptops, but also in home video game consoles, video and digital cameras, pagers, and GPS units among other electronic devices. One cell phone may contain anywhere from ten to twenty tantalum capacitors, each of which is less than half the size of a typical paperclip (if not smaller) and at least twice as delicate.[63] Tantalum is so important that the United States has stored it in the National Defense Stockpile to ensure an adequate supply for use in missiles, aircraft, nuclear reactors, and communications and weapons systems. The electronics industry now consumes over 60 percent of the world's annual production of tantalum.[64] In 2000 over two million pounds of tantalum was used in over twenty billion capacitors.[65]

Thanks to the proliferation of semiconductor chips and cell phones—the number of U.S. cell phone users grew from essentially zero in 1983 to nearly two hundred million by the end of 2004,[66] and as of 2003 over one billion cell phones were in use worldwide so by the time the high-tech bubble approached its bursting point in 2000 and 2001, coltan had become an extremely hot commodity. Between January 2000 and December 2000 the price of coltan skyrocketed, rising from about $40 a pound to $380 a pound before dropping to about $100 a pound in July 2001.[67] But by late 2004 prices had begun to rise again—mirroring the fortunes of the high-tech industry. Tantalum is not known to be particularly toxic, but its extraction from mines in central Africa has in many places devastated the adjacent communities.

In North Kivu Province, in the eastern portion of the Democratic Republic of the Congo (DRC) close to the Rwandan border, a mine collapsed in January 2002, killing at least thirty people and trapping at least as many others. In this region men have left their families and farms, teachers their schools, and women and children have left villages to work in mines where landslides—and violence—are a constant danger. Sales of the ore they mine, which has been described as the DRC's "most lucrative raw material,"[68] have helped finance the brutal fighting within the DRC and between Uganda and Rwanda in this mineral- and metal-rich region. Between 1998 and early 2005, the war, hunger, and disease have killed approximately 3.8 million people,[69], causing what has been called "one of the world's worst humanitarian crises."[70]

A United Nations Security Council report released in 2002 found that prisoners of the Rwandan Patriotic Army were being used as indentured labor—"variety of forced labor regimes" and "captive labor" are the phrases the report used—to mine coltan in the northeastern regions of the Congo.[71] Reports from journalists and NGOs describe primitive mining operations, with people standing knee and thigh deep in muddy water, working with hammers, pickaxes, and shovels, sluicing ore through plastic wash tubs and bark-stripped trees.

Hillsides, riverbeds, and land once used for grazing and farming have been bulldozed and flooded. "Entire hills and valleys have been turned into giant craters," says a report based on observations made in the Congo in 2000 and 2001.[72] Trees vital to the area's rain forest and the Mbuti people who live there have been stripped of bark and killed. The coltan-rich area of the DRC is also home to what is considered the last secure habitat for the eastern lowland gorilla. Illegal mines have been dug in the Okapi Faunal Reserve in the Ituri rain forest, and logging of the area's forests caused gorilla populations to decline an estimated 80 to 90 percent over a recent five-year period.[73]

"Convoluted" is the word that best describes the economics of the Congo's coltan trade. Local businesspeople work in partnership with foreign metals brokers. Levies and protection money are demanded by warring factions of rebel armies from both the DRC and Rwanda—some with ties to former Hutu fighters. Local government taxes are often evaded. There seem to be numerous unknown or undocumented partners profiting from and financing coltan mining and trading ventures. And this only begins to sketch the web of the central African coltan business that stretches from the Congo's rain forests to Europe, the Middle East, the United States, and China.

Among the impacts of the Congo's coltan trade are the significant and detrimental local social ramifications of an unregulated mining boom. "At the moment food is very expensive. Even in the mines we pay two or three times more, so often we return to the village with no money because the work is hard and we have to spend all we make in order to eat enough to keep going . . . Sometimes soldiers take our produce on the road. Often our employers cheat us on the sales price and give us hardly anything," one miner told an interviewer from the Pole Institute, a multicultural organization with offices in Goma in the northern Congo and in Gisenyi, Rwanda ("pole" is a Swahili word of comfort and sympathy). Nearly all the miners interviewed were concerned that abandonment of agriculture for mining was contributing to local food shortages.[74] Similar accounts have been described in numerous press reports. In response, in 2001 NGOs in Belgium began protesting the use of coltan from the Congo with the campaign slogan, "No blood on my mobile."

In truth, most coltan comes from elsewhere, though that situation may change. About 80 percent of the world's coltan is presently mined in Australia, and a number of other countries have coltan reserves, including Brazil, Canada, Mozambique, and South Africa. Nevertheless, though only about 15 percent of the coltan processed comes from Africa, that's where about four-fifths of the world's untapped coltan reserves are located. An estimated 80 percent of those reserves are in the DRC, where the geology brings the ore close to the surface, making it more easily accessible than elsewhere—political circumstances notwithstanding.[75]

The processing that turns coltan ore into the tantalum powder that is then made into capacitors takes place in China, Germany, Japan, Thailand, Kazakhstan, Russia, the United States, and a number of other countries. The companies that process the ore—among them Cabot Performance Materials, H. C. Starck, Ningxia Non-ferrous Metals, and Reading Alloys—obtain their raw materials from sources all around the world, sources that change from year to year. And the capacitors that, in turn, spark high-tech gadgets to life are manufactured literally all over the world by companies—among them Kemet, AVX, NEC, and Epcos—who sell their wares to the corporations whose names are emblazoned on our cell phones and PCs.

So in order for Motorola, Nokia, Hewlett-Packard, Sony, or any other high-tech electronics manufacturer to say where the coltan that begat the tantalum that created the capacitor that went into the cell phone came from requires conscientious and assiduous upstream documentation. But once coltan is sold into the international market—or once it leaves the mine—it becomes very difficult to trace it definitively to its source because there is no official, auditable process—analogous to the Kimberley Process established for diamonds—that would make this possible.

In late 2004 and early 2005 a shortage in the world's supply of tin ore—found in the same part of the Congo as coltan (which often occurs in the same geological formations as tin)—increased competition for control of mines near the Rwandan border and intensified the regional war.[76] Contributing significantly to the increased demand for tin ore are the new EU directives restricting the use of lead in electronics. These restrictions (mandated by the RoHS directive, or Restriction on use of Certain Hazardous Substances), have led many manufacturers to begin using tin instead of the lead-based solder that had been prevalent in circuit boards, increasing demand for tin. So, although the fierce demand for coltan has abated for a time in the Congo, the high-tech resource story in that region is far from over.

Not long after reports of extreme exploitation in the Congolese coltan trade began circulating in 2000 and 2001 and after the release of the UN report—which named individuals and companies trading in coltan from the DRC—the Brussels-based Tantalum-Niobium International Study Center announced that "civil war, plundering of national parks and exporting of minerals, diamonds and other natural resources to provide funding of militias has caused the Tantalum-Niobium International Study Center to call on its members to take care to obtain their raw materials from lawful sources. Harm, or the threat of harm, to local people, wildlife or the environment is unacceptable."[77]

At about the same time, a number of major tantalum processors, capacitor manufacturers, and companies that buy tantalum capacitors for their high-tech electronics issued statements asserting that none of the tantalum they purchase comes from miners or minerals brokers whose business supports rebel armies or organizations in the Congo. Among them was Motorola, the German company H. C. Starck[78] (a subsidiary of the Bayer Corporation), the Boston-based Cabot Corporation, and Kemet (the world's largest manufacturer of tantalum capacitors, based in South Carolina). Motorola's statement says that it purchases tantalum products from companies in Japan, Korea, and the United States that obtain "most of their tantalum from two main processors, Cabot and Starck," while Kemet "requires its suppliers to certify that their coltan ore does not originate from Congo or bordering countries."[79] This policy was also endorsed by the Electronic Components, Assemblies, and Materials Association.

But the trail of Congolese coltan is circuitous. The UN report traces coltan mined in the Congo under the auspices of an American company called Eagle Wings Resources—a subsidiary of another American company called Trinitech that is associated with a Dutch company called Chemie Pharmacie Holland (with headquarters in Ohio)—to Kazakhstan, where the mineral was processed at the Ulba Metallurgical Plant, and to the Ningxia smelter in China, as well as to H. C. Starck. The report also cites documents the United Nations believes to be false, documents that claim Mozambique as the origin of coltan that was actually mined in the Congo—a shipment of coltan that later traveled to South Africa and Thailand. Other reports gathered by the UN team follow coltan mined in the eastern Congo across the border into Uganda, where it was flown to the United Arab Emirates and then to Kazakhstan for processing. The report also traces coltan mined in the Congo moving in and out of Europe through Belgium.

Some countries listed as sources or producers of tantalum on the USGS annual fact sheets for the mineral industry, are merely transshipment points—countries (for example, the Bahamas) without tantalum mines or processing facilities—something that adds to the difficulty of independent tracking of the ore's world travels. "It's a closed little world," said Ethel Shepard, Cabot's spokesperson, when I commented on the difficulty of getting information about the source of metal ores.

Since the release of its report in 2002, the United Nations has revised its list of individuals and companies it cites as participants in illegitimate or illegal DRC coltan trading. When I contacted the Cabot Corporation—one of the companies named in the report—I was told that "all the information in the UN report was incorrect and later retracted." But the report leaves many questions unanswered and its lack of clarity raises additional questions. Among them are why certain companies were removed from the list of those doing business with what the United Nations considers illegitimate or illegal sources, why certain companies will be investigated further, and if any of the listed companies have improved their coltan-trading practices. Also unclear is why all documents related to the report are not available for independent review.[80]

Adding to the difficulty of tracing coltan back to its origins and indentifying all players in this trade are the guidelines set forth by the Organization for Economic Cooperation and Development (OECD) to prevent illegal exploitation of natural resources. Because these guidelines only apply to companies with a financial investment in or ownership of the material in question, it's possible for a company to facilitate coltan mining in the DRC without being out of compliance.

In February 2005 the DRC announced that it would "boost" its "control of coltan in the east" of the country and that by 2006 it would export only a semiprocessed form of the mineral rather than raw ore, a move presumably aimed at curtailing illegal or undocumented exports.[81] At the same time, the United Nations continues to cite control of mineral resources—including coltan and now tin—as a "significant factor" in some of the fighting in the Congo that resumed in the northeastern part of the country in late 2004.[82]

A common criticism of regulations like the European Union's RoHS directive is that banning certain materials may result in the use of alternatives that may be no better than what's being banned—or that may have new and different detrimental impacts. Some observers critical of the RoHS directive may point to the restrictions on lead in circuit boards and say that among its consequences are an increased demand for tin and the ensuing rise in prices contributing to the fighting over mines in the eastern Congo. But perhaps instead of continuing to use a known neurotoxin in the world's millions of high-tech electronics, the restriction on lead could be accompanied by an insistence on transparency throughout the entire supply chain.

Because most of the world's tantalum goes into electronics, electronics are the best source of recycled tantalum. But, writes the USGS, "although more than 60 percent of the tantalum that is consumed in the United States is in the electronics sector the amount of tantalum recovered from obsolete electronic equipment is small."[83] While the trend toward miniaturization of high-tech electronics means less tantalum is used, miniaturization also makes recycling and materials recovery more difficult. Between 1998 and 2002—when demand for and prices of tantalum were booming and many electronics shrunk dramatically in size—the amount of tantalum recycled in the United States decreased from some 35 to about 20 percent.[84] As both the Tantalum-Niobium International Study Center and the USGS note, increased cooperation between and concerted efforts by the tantalum and electronics industry are needed to improve rates of tantalum recycling.

As of late 2005 there remained no hard-and-fast way to reliably and independently trace the source of tantalum ore. The UN investigation continues, as does the conflict in the Congo. Unless all buyers of tantalum and tantalum products demand independent, legally verifiable documentation of where their ore comes from, it's likely that coltan mined under circumstances that should be considered socially and environmentally unacceptable will continue. Tantalum provides a stark reminder of the global reach of the materials that go into high-tech electronics and the challenge they present in terms of supply-chain sleuthing.

Whether the metal is copper, gold, lead, tin, or coltan, mining takes an enormous toll on the environment, uses huge amounts of resources, emits hazardous waste products, and in many places (including the United States) remains a dangerous way to earn a living. Unlike girders, beams, and plumbing, which are intended to last for decades—or jewelry that is often handed down for generations—high-tech electronics last only a few years. So it seems particularly profligate to send the metals in high-tech electronics to the dump, especially when it's far less resource intensive to recover and reuse metals than it is to mine and refine ore. Because high-tech electronics are so complex, it may never be possible to utterly simplify their manufacturing process, but it should be possible to reduce waste products and close production loops as shown by efforts being made to lessen the environmental impacts of producing silicon wafers. Yet metals production and turning crystalline silica into the foundation of semiconductors is only the beginning of the high-tech production process—a chemical-intensive process whose materials have been linked to a number of disturbing health effects.

CHAPTER THREE

Producing High Tech

The Environmental Impact

In many ways the revelations about high-tech's environmental impacts begin underground, not in a mine, but in the soil and water that form the ordinary terra firma of our everyday lives. Which is fitting, because like the electrical signals speeding through a circuit board, many of the environmental impacts of high technology are virtually invisible and therefore easy to ignore—at least for a time. So it's possible to say, as California State Senator Byron Sher, whose district encompasses much of Silicon Valley, described it to me that public awareness of the environmental impacts of the high-tech industry began with LUST (the professional acronym for Leaking Underground Storage Tanks)—LUST in the heart of the red hot center of the high-tech universe.

LUST IN SILICON VALLEY

"By the mid- to late 1970s," write the authors of *Fire in the Valley: The Making of the Personal Computer*, "the fire of invention burned brightly in Silicon Valley, fueled by a unique environment of universities and electronics and semiconductor firms."[1] The geography of Silicon Valley is roughly defined by the cities of San Jose, Mountain View, Sunnyvale, and Cupertino to the south and Menlo Park and Palo Alto to the north, communities that sprawl south from San Francisco Bay across the Santa Clara Valley, known in its agricultural heyday as the Valley of Heart's Delight. By the end of the 1970s, the electronics industry had thoroughly overtaken what had been one of the world's premier fruit-growing regions. Where peaches, cherries, and apricot trees once bloomed, soil was churned to make room for burgeoning semiconductor, computer, and related high-tech companies.

Advanced Micro Devices, Fairchild Semiconductor, Hewlett-Packard, IBM, Intel, National Semiconductor, NEC, and Raytheon are just a few of the better-known companies that set up shop in the area, turning out the complex bits of circuitry that power the brains of high-tech electronics and that are, in effect, the currency of the Information Age. As I drive through acres of office parks, strip malls, and residential developments in March 2004, I find it hard to imagine what the valley looked like when there were more orchards than cement. The area's sleek commercial buildings punctuated by tidy suburban shrubbery could house anything. Very few have any visible duct work, smokestacks, chimneys, or any other hint of industrial manufacturing. There are no signs to alert a visitor that Santa Clara County has, thanks largely to high tech, more Superfund sites than any other U.S. county. From the outside, it's as hard to envision what goes into making the high-tech electronics as it is to picture this freeway-laced landscape awash in fruit blossoms.

Compounding this difficulty is that almost everything we read or hear about high-tech electronics focuses on what these devices do, and on the businesses that design and sell them, rather than on the materials and manufacturing processes that create them. And unless you are or were a resident of a community where the semiconductor industry put down early roots or have worked in a semiconductor fabrication plant, you're unlikely to be aware of any possible connection between local water or air quality and the computer on your desk.

While Silicon Valley can be considered the birthplace of high technology, it's far from the only place where high-tech manufacturing takes place. Rust Belt communities in the Northeast, rapidly expanding cities in the Southwest, the Pacific Northwest, and Texas, as well as cities in China,

Malaysia, the Philippines, Indonesia, Japan, Ireland, and Israel are but a handful of the places home to semiconductor and other high-tech industrial plants. The global reach of this industry is an important factor to consider when evaluating its environmental impacts or ecological footprint. While environmentally minded grocery shoppers can choose to buy local produce, hometown baked goods, and sometimes milk and cheese from nearby dairies, high-tech consumers don't really have an equivalent option. The whole industry is designed to take advantage of efficiencies created by using a global supply chain. This often includes operating in places like China, where labor and other business costs may be lower than in the United States or western Europe, and it means working on a global clock, which enables production lines to run twenty-four hours a day, seven days a week.

As I learned by visiting the Intel Museum in Santa Clara, "An Intel chip has likely traveled the world before reaching your computer. A processor designed in Portland, Oregon, may be fabricated in Ireland, and then assembled in Costa Rica." Not only are the various local impacts of high-tech manufacturing plants part of these products' ecological footprint, then, so are the impacts of transportation. But because the price of high-tech electronics continues to decline relative to the equipment's capabilities, it's difficult from a consumer's perspective to realize that creating these products has costs that extend far beyond what one sees on the vendor's price tag. Which brings us back to LUST, those scores of underground tanks in Silicon Valley.

To understand the connection between what was stored in these tanks and the water quality in Mountain View, California, or the indoor air quality in Endicott, New York, one needs a basic sense of what semiconductors actually are and how they're made. For the materials in those tanks were used to make the microchips that laid the groundwork for what enables your computer to perform a Google search or your iPod to play a song.

THE MAKING OF THE SEMICONDUCTOR

Semiconductor production, as discussed earlier, involves numerous chemicals and chemical compounds, solvents, and metals, many of which are hazardous and toxic. The waste from this process has, over the years, included nitric acid, barium, chromium, copper, and nickel, and solvents (which were originally stored in those underground tanks), among them methanol, glycol ethers, and ammonia. All of these substances have often ended up—one way or another—in local rivers, streams, and aquifers, often in great volume where high-tech manufacture takes place.

‖█ ‖‖ █

Semiconductors, microchips, chips, or integrated circuits (as they are also called) are the thin slices of highly polished wafer onto which are etched the intricate transistor pathways that transform electrical current into a digital source and repository of language, information, and numerical calculation. Transistors are the microscopic switches that control the flow of electricity that makes a computer compute, and the microchip forms the brains of the computer. The part of the computer that contains the semiconductor is called a processor or microprocessor. "Micro" is the operative word, for chips of the current generation are so small that it's virtually impossible to examine a single chip's circuitry without magnification. For example, one of Intel's Pentium 4 processors contains fifty-five million transistors linked by circuitry only 0.13 microns wide[2]—a micron is one-millionth of a meter, which makes the processor a little bigger than one-tenth of a millionth of a meter, or about a thousand times thinner than one human hair. The etched wafer patterns are so tiny and complex they make snowflakes look like manhole covers.

Given their scale, semiconductors use ionically charged chemicals rather than mechanical constructions to create a network of switches. These switches can signal "in/out" or "on/off," and thus create the fundamental logic functions that enable computers to "select this *and* this but *not* this,"[3] and which form the basis of the digital universe.

Each individual chip's pattern of switches translates into a code of bits and bytes programmed to register as units of language, mathematics, or graphics. Eight bits make 1 byte, and 256 bites equal 1 hertz (the computer I'm using, for example, has a 900-megahertz processor). The chemistry of the semiconductor also allows the chip to remember specific patterns of code and therefore to remember what creates, say, the letter "a" or a string of letters or functional directions.

One way to think about these switches and signals is to imagine those simple circuits you may have made as a child, with wires and batteries—where closing a switch rang a buzzer or lit a bulb—ratcheted up to the *n*th degree of complexity and then shrunk down to an infinitesimal scale. Instead of wires and mechanical levers to make the pathways of signals and switches that form a microprocessor, chemical developers carve a designated pattern onto a slice of semiconductor material, a substance that can be electronically or chemically modified to act as either a conductor or insulator of electricity.

Silicon is by far the most commonly used semiconductor wafer material, but there are others, including germanium, gallium arsenide, and indium phosphide. The wafer material determines what other materials and chemicals will be used to create the semiconductor, and also what kinds of waste products will be generated.

Before the process of creating the circuit patterns begins on silicon wafers, an invisibly thin layer of silicon dioxide is grown on the wafer, which is made of highly polished crystalline silicon. The silicon dioxide layer—where the circuitry will be etched—is created by exposing the silicon to high heat and a gas, usually oxygen. Chlorine—which may be in the form of chlorine gas, hydrochloric acid, trichloroethylene, or trichloroethane—is sometimes used to modify this layer to suit a particular chip's specifications.[4]

Next the wafer is coated in a photoresist—a finely calibrated and chemically sophisticated combination of a light-sensitive polymer resin (itself made up of numerous ingredients) and a solvent that makes up a large part of the mixture. These photoresist chemicals are continually evolving. They differ depending on what kind of light (ultraviolet or gamma rays, for example) is used to etch away the resist and on how fine the desired geometry of the circuitry is.

Some changes in these etching-process compounds have been made to solve previously recognized safety issues. As a class of chemical compounds, solvents (excepting water) range from mildly to very toxic. Those that contain aromatic compounds (molecules that contain benzene rings) and chlorinated hydrocarbons are considered to be the most hazardous. Some are known to cause cancer and numerous other health problems, among them skin rashes, lung disease, anemia, kidney and liver damage, and impaired immune system function. Some solvents are reported to adversely affect fetal development.[5]

Semiconductor Equipment and Materials International (SEMI), an organization of companies that supply material and equipment to the semiconductor industry, says that solvents used in photoresist formulas that were developed in the 1970s, and which posed problems of flammability, emissions, and disposal, have been phased out of production. For similar reasons, notes SEMI, since the 1990s, there has been a move to substitute aqueous or water-based solvents for organic solvents (a class of compounds that includes ethylene glycol, trichloroethylene, and trichloroethane) in various parts of the semiconductor fabrication process.[6]

Although the chemicals and fabrication processes have evolved at a rapid pace, the legacy of chemicals used in the 1970s and '80s—chief among these, trichloroethylene and trichloroethane—persists. In the first decades of semiconductor fabrication, these chemicals were stored in underground storage tanks that often leaked, and spills occurred. Such contamination lingers for decades, posing health risks to communities not only in Silicon Valley but in many other places where high-tech industry settled. But you won't find that part of the story in any of the technology museums or in any of the guides to how a computer works.

What you will learn from the interactive museum models and diagrams is that after the photoresist is applied, the circuitry pattern is etched into the silicon dioxide by exposing the wafer to ultraviolet light through something called a mask. Think of a stencil made, not of paper, but of chrome-covered quartz. When exposed to ultraviolet light the photoresist becomes gooey and soluble. The mask is then removed and the rest of the photoresist is dissolved with solvents before another layer of circuitry is etched onto the chip. There are both "wet" and "dry" etching processes. In wet etching a liquid acid is typically used, while in dry etching bromine, chlorine, fluorine, or iodine-based gases are used—halogenated compounds that can release hazardous air pollutants that may include chlorine or hydrochloric acid.

"Chemicals and gases are used throughout the chip-making process. Some, like hexamethyldisilazane, are complex and difficult to pronounce.

Others, such as boron, are simple elements found in the Periodic Table of the Elements," reads Intel's description of the chip-making process.[7] The materials used in the etching process make a daunting list of chemical compounds. Among them—and this is just a sampling—are acids, including hydrofluoric, nitric, phosphoric, and sulfuric acid, as well as ammonia, fluoride, sodium hydroxide, isopropyl alcohol, and methyl-3-methoxypropionate, tetramethylammonium hydroxide, and hydroxyl monoethanolamine, along with acetone, chromium trioxide, methyl ethyl ketone, methyl alcohol, and xylene. Altogether, one individual semiconductor fabrication plant may use as many as five hundred to a thousand different chemicals.[8]

The process of coating a wafer with photoresist, laying down a mask, etching and removing the remaining photoresist is repeated many times; a single chip may have twenty or more such layers etched onto its silicon base. After each layer is etched, that layer is bombarded with gases in a laserlike process called doping, which sets up the precise flow of electrical charges desired for the chip.

The gases used in this doping process are often boron and phosphorus, but can also include argon, arsine, silane, and phosphine, all of which are highly toxic. Other dopants include arsenic, antimony, beryllium, chromium, and selenium. Some are extremely flammable and can ignite when brought into contact with oxygen. The International Labour Organization (ILO) calls dopants "potentially the most hazardous group of chemical used in electronics." If a leak or rupture occurs, releasing one of these substances—phospine, arsine, or a borane, for example—"the whole factory and surrounding community can be affected with many cases of serious harm and sudden death."[9]

To connect the layers of circuitry, atoms of metal—commonly of copper and aluminum, but gold, platinum, nickel, cobalt, tungsten, titanium, or molybdenum may also be used—are deposited on the chip. The scale that's being worked on at this stage, say my guides at one of Intel's plants, is like aiming a tennis ball at the earth from the moon. After this step, a layer of electrically charged polysilicon is placed in the middle of the chip. This layer will be used to carry electrical signals sent from an outside power source. After the circuitry fabrication is complete, the wafer is polished to make it as smooth and flat as possible, often with a mixture of water and silica. When all the work on a single wafer is complete, the individual chips are cut apart with a diamond blade saw. One three-hundred-millimeter wafer contains hundreds of chips that are made at the same time, and often two dozen or more such wafers are processed in a single batch.

In a step called packaging, the tiny chip is then placed into a piece of plastic or ceramic with tiny wires—leads, often minute threads of gold—that extend from the chip to connect it to other parts of the circuit board.[10] Solder and adhesives are used in this part of the process. Traditionally, the solder used has been lead, but in an effort to eliminate that well-known neurotoxin from circuit boards (which will be an EU requirement starting in 2006), semiconductor manufacturers are developing lead-free solders, many using tin and silver.

According to my guide at Intel's Jones Farm Campus in Hillsboro, Oregon, chip production involves over 250 steps—the photolithography process alone involves about two dozen steps. And between 200 and 250 people work in each of Intel's Chandler, Arizona, fabs, I'm told, in twelve-hour shifts that run twenty-four hours a day, seven days a week; but much of the process is actually automated. Replications of what happens in Chandler are carried out in semiconductor fabrication plants in locations as diverse as Chengdu, China, the Philippines, Costa Rica, Ireland, Israel, India, and Malaysia (and these are just one company's plants).

||| ||| |

With even this rudimentary knowledge of semiconductor production—and its many chemicals—one can begin to understand how the manufacture of one two-gram microchip (about 0.07 ounces, less than the weight of one teaspoon of milk[11]) can generate almost 26 kilograms (57.2 pounds, or about thirteen hundred times its weight) of waste, some of it highly toxic—and this amount doesn't include air emissions.[12] Back in 1998 when semi-conductors were larger and manufacturing less efficient than today, fabri-cating chips used in a typical desktop computer generated about ninety pounds of waste, about 8 percent of which was toxic.[13] Many improvements in materials efficiencies and environmental impacts have been made in the intervening years—of which more later—but the semiconductor industry continues to be one of the most chemically intensive ever.[14]

MEASURING FOOTPRINTS

Measuring the ecological footprint of semiconductor manufacture or a finished piece of high-tech electronics—whether a mammoth flat-screen television or a palm-sized cell phone—involves what's called doing a life-cycle analysis. This means looking at all the materials and processes that go into making a product, as well as the waste products of that manufacturing process, and then summing up these findings in a meaningful way. To better understand how one goes about doing such a calculation and how one uses the results, I spoke to Eric Williams at United Nations University in Tokyo, who is something of an expert on life-cycle analyses of high-tech products and their applications.[15]

"These are complicated analyses," Williams tells me. To begin with, you need reliable and accurate information on the quantities of materials, energy, and water that go into producing a semiconductor or silicon wafer. Then, for high tech, there's also the issue of how to account for the ecological effect of ever-escalating product volume. High-tech manufacturers "keep trying to get across the fact that the efficiency per product is going up, so we don't need to worry, things are getting better," Williams comments. "In any industry technological progress induces efficiency. But it also induces growth. So if one doesn't keep pace, there will not be an overall improvement."

Life-cycle analyses, while they may appear to be dealing simply with cold, hard data, can also be controversial, as Williams and his colleagues discovered after publishing their analysis of a microchip in the journal *Environmental Science and Technology*. One response came from Farhang Shadman, director of NSF/SRC engineering research at the University of Arizona's Center for Environmentally Benign Semiconductor Research, and Terrance J. McManus, Intel fellow and director of Intel's environmental health and safety technologies:

We feel that this study's conclusion is itself misleading because using weight as the basis for comparison is arbitrary, non-scientific and inaccurate. Product weight is a metric that does not account for performance, utility and benefits of these products . . . By not including the value and benefits derived from high-technology products, one reaches the erroneous conclusion that products such as semiconductors, pharmaceuticals and software present a disproportionately higher environmental burden than products of other industries.

To which Williams and his colleagues responded:

Our point was, and is, simply that the amount of materials used to manu-facture a computer chip these days is hundreds, if not thousands of times greater than the quantity actually embodied in the chip. This makes the weight of the chip a misleading indicator of the amount of materials used, and it means that people like Alan Greenspan and Frances Cairncross who have cited microelectronics as an example of radical "dematerialization" have misunderstood the situation . . .

While not the topic of our article, we agree with the letter writers that the semiconductor industry has been improving its environmental perform-ance. We have yet to see, however, that this progress has been sufficient to counteract growth in demand so as to realize net decreases in material and energy used by the industry. According to US Census statistics and consult-ing firm data, wafer, chemical and energy use of the semiconductor indus-try increased 6–10% annually in the late 90s. While this rate of increase is less than the economic growth of the industry (about 15% on average), it is still substantial. The evolution of materials intensity at the product level is an interesting question as well. While the smaller feature size of newer genera-tion chips could imply less materials use per transistor, the increased com-plexity of processes, the requirement for ever-declining defect densities, and the need for purer starting materials have the opposite effect: the ratio of indi-rect to direct materials consumption may actually be increasing.[16]

When I asked Williams about this exchange, he explained how careful one has to be in a life-cycle analysis not to compare apples to oranges. One of the most important considerations of a life-cycle analysis is deciding how far up and down the process and supply chain you're going to go—how many sources for materials and manufacturing processes you're going to include. "In practice, you can never do them all, but you do as many as you can," he says. "But you have to define the system boundaries and explain what's in and what's out." This is essential for establishing a basis of comparison.

For example, if you're going to compare the amount of water used in semiconductor manufacture to the amount of water used to produce, say, an egg, a pair of blue jeans, or the Sunday *New York Times*, you have to know how far back up the source material chain the analysis goes. Is the water required to grow the grain that fed the hen included in the analysis, or just the water the hen drank while the egg was growing? Is the water used in growing and processing the cotton, weaving the denim, making the dye and the thread, and running the factories where all this took place—as well as the sewing of jeans—included? Are we talking about simply printing the newspaper or about producing the newsprint, growing and logging the forest, and operating the buildings where all this takes place as well?

When I asked Intel how much water is used in their fabrication of a semiconductor, I was told, "It depends on the complexity of the chip but we estimate about 3.5 gallons per Pentium 4 chip." As a point of comparison, Intel said that it takes 120 gallons to produce one egg. Curious about this figure, I checked in with the American Egg Board, information headquarters for the "incredible edible egg." They referred me to Dr. Jim Arthur of Hi-Line International, an egg producer in West Des Moines, Iowa.[17] According to Dr. Arthur, a flock of laying hens needs 4 to 5 gallons per hundred birds, and the flock lays, on average, between eighty to ninety eggs a day. Dr. Arthur's calculations and experience suggest that a hen would need between six and eight ounces of water to lay eight to nine eggs. "The 120-gallon figure would have to include water for something other than what the hen consumes directly," Dr. Arthur said. Intel's 3.5-gallon figure is for water used directly within the semiconductor lab and doesn't include water required for what the life-cycle analysts call "inputs," which in the case of the hen might include the grain she ate during an egg-laying cycle.

"This is the twenty-first century. You think we'd be on top of this" kind of study, says Williams. "But I don't think we are."

One of the reasons it's so difficult to get a grip on the materials involved in semiconductor manufacture is the frenetic pace with which new materials are developed and introduced. "The industry's push to keep up with . . . [Moore's] law* with each generation of chip has always led to manufacturing and materials changes, but in recent years these changes have been profound. 'There were many generations over which one or maybe two new materials were introduced,' says Pam Mattimore, president of Air Products Electronic Chemicals.[18] 'Now we're talking multiple materials, and they are all piling up on top of each other,'" wrote Michael McCoy in a cover story for *Chemical and Engineering News*[19] "Getting in early really sets the market," said another industry executive. "Even if you have a great product, if you come in late it may not matter."[20]

"The need to develop and deliver new process technologies at a breakneck pace necessitates that the technologists focus almost exclusively on developing a process with the efficiency and robustness to deliver maximum yield," wrote Jay M. Dietrich of IBM in a paper presented at the 2004 IEEE International Symposium. "Optimal utilization of input commodities—chemicals, water, power, exhaust—receives only cursory consideration in the drive to establish a workable process for the manufacturing floor."[21] If you start the high-tech clock in 1954, with Texas Instruments' first commercial production of silicon transistors and IBM's first mass-produced computer,[22] the industry has behind it about fifty years' worth of using an almost unfathomable amount of "input commodities" and creating a corresponding volume of waste.

Many of the chemicals used to make semiconductors are so new that there are no comprehensive health and safety analyses available. And some of the older chemicals, including some of the toxics that have been phased out, like trichloroethylene and ethylene glycol, continue to be a cause for concern and controversy, with health impacts that are not yet fully understood.[23] "It used to be six to eight years from research and development to product, now it's two to three years," Ted Smith, founder, senior strate-

---

* Moore's Law—named for and first expressed by Gordon Moore, a founder of Intel—that the computing capacity of semiconductors would double every twelve to eighteen months.

Planned
Obsolesence

gist, and former executive director of the Silicon Valley Toxics Coalition told me, recounting a conversation he'd had with an occupational health doctor at IBM. "The people who in theory are in charge of health and environment are getting squeezed out of the process, so that their job ends up being a job of dealing with the consequences."[24]

Environmental advocates like Smith are not the only ones with this point of view. "Conventional methods used to address global environmental programs have only been moderately successful," write Patrick D. Eagan and his colleagues at the University of Wisconsin. These methods, he adds, "are reactive; the pollution is created and then treated to prevent escape to the environment."[25] To be truly successful, however, methods have to be proactive: one has to know what the potential pollutants are—and have knowledge of their behavior and impacts—before they enter the waste stream or become part of products that go into our homes and offices, later ending up in landfills and incinerators either at home or abroad.

Further complicating accurate analysis and assessment of health consequences is the difficulty of obtaining all the necessary information. There may be anywhere from two hundred to three hundred different ingredients involved in making a semiconductor, materials that vary depending on the type of semiconductor being made and the fabrication process. As Gary Niekerk, Intel's environmental health and safety external affairs manager explained, the specific materials used in chip production are generally proprietary to the chip's manufacturer.[26] A semiconductor manufacturer will reveal the general categories of material used in chip production, but it's hard—if not impossible—to get an exact list of what goes into making a particular manufacturer's chip.

In an effort to address this dilemma, particularly as it relates to disposal and recycling of high-tech products, members of the Electronics Industries Alliance (EIA) have created a "product material declaration program." The program was prompted in part by the European Union's Restriction on Hazardous Substances (RoHS) directive, which will bar use of certain toxics in electronics, and its Waste Electrical and Electronic Equipment (WEEE) directive, which mandates e-waste recycling in the European Union. When these regulations go into effect, all high-tech manufacturers selling goods in the European Union will be required to supply recyclers with a materials list for their products in order to facilitate efficient and safe recycling.

Meanwhile, independent of WEEE and RoHS, many businesses, institutions, local governments, and organizations in the United States and elsewhere have begun to incorporate materials declarations into their purchasing policies. The EIA began their experimental materials declaration program in 2001, and now Advanced Micro Devices, Intel, and other manufacturers of both components and consumer electronics make information available, within the bounds of what they consider proprietary.

"WE CALL THESE PIRANHA BATHS BECAUSE THEY EAT THINGS"

While the information available publicly may not be sufficiently detailed for researchers like Eric Williams to analyze meaningfully, the major semiconductor and high-tech electronics manufacturers all publish environmental health and safety reports of their own. These assessments report on the companies' use of certain toxic chemicals, their release of hazardous air pollutants, use of water, and worker health and safety records, at both domestic and overseas facilities. And in the thoroughly globalized marketplace of high tech, high-tech manufacturers are paying increasing attention to the practices of their entire supply chain.

Intel, easily the United States' best-known and the world's largest semiconductor manufacturer, has been publishing its environmental health and safety report since 1994. Founded in 1968 and admired for both its products and its competitive business strategies, Intel initiated its health and safety reviews in 1982 and its environmental reviews in 1984, issuing its first safety and environmental policies in 1985.[27]

"We have a set of worldwide standards for pollution prevention and resource conservation," says Philip Trowbridge, who works on environmental health and safety issues at Advanced Micro Devices (AMD).[28] "We have developed goals for each site and plans to achieve those goals," he tells me, speaking from his office in Austin, Texas. Founded in 1969, the company is now one of the leading U.S. semiconductor manufacturers with operations in Thailand, China, Malaysia, Japan, Germany, and the United States.

Like Intel, AMD has had an environmental health and safety department since the early 1980s, Trowbridge tells me. The company developed its current environmental standards and began publishing an environmental health and safety report around 1996. It was in the following few years that thinking about sustainability in all of their business practices really came into play, adds Trowbridge. What prompted this?

"A lot of it has to do with what was going to be coming down from Europe," Trowbridge says, alluding to the WEEE and RoHS directives. Also playing a role have been the European Union's commitment to international agreements on toxics and greenhouse gas emissions (the Stockholm agreement on persistent organic pollutants and the Kyoto Protocol on climate change, for example) and proposed EU policies on use of chemicals in consumer products (the REACH legislation, Registration, Evaluation and Authorization of Chemicals). There were also "questions from investor groups," says Trowbridge. What happens in Europe and elsewhere around the world is very important to AMD because the company earns nearly 80 percent of its revenue internationally.[29]

And, says Trowbridge, "It's no secret that we have three Superfund sites in Silicon Valley," referring to the AMD plants identified in the investigation of leaking underground storage tanks in Santa Clara County. "We do have that skeleton in our closet," he continues, "but we're getting to the end of life on that," meaning that the cleanup process is nearing completion.

▌▐▐▌▐

A look at any major high-tech manufacturer's environmental health and safety reports will show a net improvement in environmental efficiencies. There are some fluctuations, but overall, less per-unit water is being used, less energy, a smaller volume of chemicals, and there are fewer releases of toxics to air and water and less emission of greenhouse and ozone-depleting gases. Without taking away from these achievements, it's worth noting that some of these net improvements have come about because more product is being made with the same amount of resources or because manufacturing facilities have been closed or sold.

As IBM notes concerning its use of chemicals for which the U.S. Environmental Protection Agency requires reporting, "From 2002 to 2003, IBM achieved a 20.5 percent decrease of the total quantities covered by both SARA [Superfund Amendments and Reauthorization Act] and PPA [Pollution Prevention Act] worldwide to a total of 4,202 metric tons. The majority of this reduction was the result of the divestiture of some operations, but pollution prevention initiatives also contributed to the perform- ance." IBM has a similar comment about its production of hazardous waste: "From 2002 to 2003, IBM's total hazardous waste decreased by 2,704 met- ric tons or 19.8 percent. Though the sale of some operations, primarily hard disk drive manufacturing, accounted for the majority of the reduction, pol- lution prevention actions also contributed. IBM recycled approximately 46.2 percent of the hazardous waste it generated in 2003."[30] Comparable obser- vations can be made of most other major high-tech manufacturers.

"We work closely with our product groups, with suppliers and research and development folks to see where we can have our largest impact," says Trowbridge. "And that's basically in two areas, chip design and resource effi- ciency. How do we design a faster, better product that also takes into account chemicals safety and reduces the amount of hazardous waste? Ultimately it's a combination of a smaller environmental footprint. That's the goal."

"Global-warming gases are an example," he continues. "We're look- ing at new manufacturing techniques that will reduce the use of global- warming gases. We're not being required to do this, but it's the right thing to do. It's a combination of us pushing and saying, we can do this, and our technologists saying, yes we can. AMD also aims to reduce the amount of waste we send to landfill by one thousand metric tons and to reduce our current manufacturing use of water and greenhouse gas emissions by 40 percent by 2007," says Trowbridge, and by that deadline also to reduce energy use by 30 percent and emissions of perfluorocar- bons by 50 percent.

Perfluorocarbons (PFCs) are used in semiconductor etching and are one of the greenhouse gases to be curtailed under the Kyoto Protocol.

PFCs, says the U.S. Department of Energy, are also "characterized by long atmospheric lifetimes (up to 50,000 years); hence . . . they are essentially permanent additions to the atmosphere."[31] The semiconductor industry began using PFCs as a substitute for ozone-depleting chlorofluorocarbons (CFCs) that, as the Semiconductor Industry Association (SIA) puts it, lost "through an international regulatory edict."[32] Throughout the 1990s, rapid expansion of the semiconductor industry caused a corresponding rapid growth in the use of PFCs. Their use and emissions peaked in 1999, according to the Department of Energy,[33] and since then the industry has been working to curb both use and release of PFCs. The U.S. semiconductor industry is now working under the terms of an agreement with the EPA in which SIA members have agreed to reduce PFC emissions by 2010 to 10 percent lower than those of 1995.[34]

Some changes that improve environmental impacts are very simple, Trowbridge explains. "For example, the baths used for etching wafers. We call these piranha baths because they eat things. Can we use less acid per wafer? How long are these baths effective for? Can we exchange the contents of these baths less frequently and reduce the amount of water and chemicals we use? Yes, we can."

But some changes are more challenging, such as finding safe alternatives to toxic chemicals used in the photolithography and photoresist process. One of the chemicals that has been used in photoresists, explains Trowbridge, is something called perfluorooctane sulfonate,[35] a chemical compound that turns out to be a persistent organic pollutant. "We've been successful in eliminating it from the developing process," Trowbridge tells me, "but the challenge has been to find a new photoresist that doesn't use it."

A TOXIC PERSISTS: ONE EXAMPLE

Trowbridge was the first in the semiconductor industry to mention perfluorooctane sulfonate to me by name. Curious, I looked it up and discovered that it's in the class of chemicals that were manufactured by 3M for use in Scotchguard, among other consumer products and applications. Perfluorooctane sulfonate (PFOS) received a flurry of attention around 2000, when studies of its effects on animal and human health showed that it was bioaccumulative (meaning that it accumulates in body tissue, especially in fat where it can linger for years), persistent, and toxic to mammals. In animal studies PFOS has been determined to be toxic to the liver, to cause damage to the thyroid, and to cause other health problems. Epidemiologic studies indicated an association between exposure to PFOS and bladder cancer in humans. PFOS has been found in effluents and sludge from sewage treatment plants and in marine mammals, fish, and other wildlife.

In 2000 the EPA published a notice in the *Federal Register* stating its intention to curb "and, if necessary, to prohibit" the use of PFOS and related chemicals under provisions of the Toxic Substances Control Act. "EPA believes that this action is necessary because the chemical substances included in this proposed rule may be hazardous to human health and the environment," wrote the agency.[36] In response, 3M—the major U.S. producer of these chemicals—announced its intention to phase the chemicals out of production.

The EPA proposal was then opened for public comment. The comments delivered to the EPA included what the Semiconductor Industry Association called "a joint effort" by "SIA and photoresist suppliers" to persuade the EPA "that these chemicals are used in small quantities and are soundly managed, posing no risk to worker health or the environment."[37] The upshot was industry success "in retaining the use of perfluorooctyl sulfonates (PFOS) and perfluoroalkyl sulfonates (PFAS) in leading-edge photoresists." Consequently, in 2002, the EPA published its rule on these chemicals that includes exemptions for use in semiconductor photolithography. Part of the industry's argument to the EPA was that current processes that use PFOS have virtually eliminated occupational exposure and have dramatically reduced discharge to the environment via wastewater or other conduits. The industry also pointed out the lack of viable alternatives and argued that, while it was committed to continued efforts to reduce the use of PFOS, these chemicals were vital to its commercial success and to "its technological contributions to national security."[38]

An interesting footnote to this story, which highlights the difficulty of tracking toxic chemical use, is that until the semiconductor industry submitted its comments arguing against the EPA's original proposal, the EPA didn't know that PFOS and PFAS were being obtained from sources other than 3M or from sources outside the United States. The semiconductor industry had in fact been importing PFOS, albeit at quantities below the hazardous substance reporting limit of 4,400 pounds per year.[39]

Given the continued concern about theses chemicals' toxicity and persistence—"and their apparent widespread occurrence"[40]—in 2004 the Organization for Economic Cooperation and Development sent its members a questionnaire on the use and production of PFOS and related chemicals, to further assess impacts and plans for continued use of these chemicals. Meanwhile, PFOS and their chemical cousins are out there in the environment, lurking in many products in our homes and offices and perhaps percolating through our bloodstreams. If we're lucky, we may never suffer their effects. Yet that this toxic substance could be so widely used, and that its use could persist despite its apparent hazards and without adequate oversight, though, seems to illustrate the flaws in our system of monitoring such chemicals.

TOXICS AND THE RIGHT TO KNOW

In the United States, semiconductor and other high-tech electronics manufacturers are required to report on the use of toxic chemicals and hazardous emissions when such toxics and emissions fall within the guidelines of the EPA's Toxics Release Inventory. This database—based on self-reporting by the businesses that use or produce the chemicals in question—tracks releases of toxics to air, water, and solid waste disposal sites. But the EPA's reporting requirements cover only a finite list of about 650 chemicals and apply only when a facility manufactures or processes more than 25,000 pounds or otherwise uses 10,000 pounds of any listed chemical each year, although there are some chemicals with much lower thresholds.[41] That means these guidelines don't include some of the toxic chemicals used or released in the semiconductor and circuit board production process, because the chemicals are not on the EPA's list or because the chemicals are not present in sufficiently large quantities.

Another problem in tracking environmental and health impacts of commercially used chemicals is that the Toxic Substances Control Act (TOSCA)—passed in 1976—required registration of chemicals that were introduced *after* implementation of the act in 1979. Some 80 percent of the chemicals used today were introduced before 1979 and don't, until further determinations are made, fall under any TOSCA restrictions. This leaves thousands of industrial chemicals untested for impacts on human health. Many of these are used in semiconductor and other high-tech manufacturing. The EPA's toxics reporting requirements are also designed to protect proprietary information.

While all the major high-tech manufacturers include their overseas facilities in health and safety reports, the EPA's reporting requirements address only what happens on-site at individual U.S.-based manufacturing locations. This could allow companies to simply move toxics overseas, but, claims AMD's Trowbridge, echoing what I hear from other high-tech companies, they implement the same environmental health and safety standards throughout the corporation. Those standards are often "above and beyond in-country standards."

The European Union's impending reporting standards for toxic chemicals promise to be more stringent, comprehensive, and precautionary[†] than those now in place in the United States. Other countries also have reporting requirements for toxics, but many of these programs—for example, those in Australia, Canada, and Japan—are more limited than the U.S. standards. And the United States and the chemicals industry have been lobbying hard to curtail or derail the European Union's proposed REACH legislation, which will require testing of chemicals before they're produced commercially or used in manufacturing and will remove from the market those now in use that cause birth defects or are known carcinogens. Such legislation would affect many industries, but it might prove particularly cumbersome to the high-tech industry, which is one of the most chemical-intensive ever.

---

† The European Union has incorporated the precautionary principle into its environmental policies, which means they will act on sound evidence or indication that a substance is harmful rather than waiting for confirmed proof, which is the legal standard in the United States.

WATER AND ENERGY

In addition to chemical compounds—the metals, gases, solvents, etchants, chemical polymers, resins, and related materials—semiconductor fabrication requires substantial amounts of energy and water. In 1997, the production of one six-inch silicon wafer required, on average, 2,275 gallons of deionized—that is, superclean—water.[42] Efficiencies have improved since then, especially on a per-chip basis and in terms of recycling water, but research published in 2002 found that 32,000 grams of water were required to produce one 2-gram microchip, hundreds of which are produced on a single wafer.[43]

In a number of places the rapid growth of the high-tech industry that occurred in the mid-1990s took up significant gulps of the local aquifer. For example, in the mid-1990s, Albuquerque, New Mexico, the third driest state in the nation, was home to a number of major high-tech manufacturers, among them Honeywell, Intel, Motorola, Philips, and Sumimoto. In 1995 these companies together used over 1.7 billion gallons of water.[44] That same year, in Austin, Texas, AMD, IBM, Sematech, Motorola, Texas Instruments, and other high-tech manufacturers used over 300 million gallons of water a month.[45]

Between 1999 and 2003 Intel's use of water at all their plants worldwide grew from about 5 billion gallons to about 6 billion gallons, or enough to cover the island of Manhattan in water one foot deep.‡ AMD's worldwide water use during the same period decreased slightly from 7.06 billion liters (1.86 billion gallons) to 5.91 billion liters (1.56 billion gallons), in part because of plant closures. Both companies, however, point to what they call the "normalized production basis" of their water use, meaning how much water they used for the amount of product manufactured. Intel's actual water use, which increased 4 percent between 2002 and 2003, was 9 percent more efficient on a per-semiconductor basis. Similarly, while AMD's actual use of water dropped from 6.39 billion liters (1.69 billion gallons) in 2002 to 5.91 billion liters (1.56 billion gallons) in 2003, the normalized amount of

---

‡ One acre foot of water—enough water to cover one acre with one foot deep of water—is about 326,000 gallons. Thus 6 billion gallons is roughly 15,000 acre feet of water, while the area of Manhattan is 14,478 acres.

water dropped more dramatically, falling from 11.07 billion liters (2.92 billion gallons) in 2002 to 7.85 billion liters (2.97 billion gallons) in 2003.

When I visited Intel's Chandler, Arizona, site—the 705-acre Ocotillo Campus—Senior Environmental Engineer Len Drago explained to me how the company is reclaiming 500 million gallons of wastewater a year and recycling some 400 million gallons internally.[46] This represents, said Drago, about 72 percent of the water the company uses there. As part of these efforts, Intel has also helped the city of Chandler finance what Drago describes as "a reverse osmosis system that treats wastewater to drinking-water standards and then reinjects the clean water back into the ground to directly regenerate the groundwater supply."

Treating wastewater at a semiconductor lab is complex, and at an Intel plant it is usually treated on-site before it goes to a municipal treatment plant, explains Gary Niekerk, Intel's Environmental Health and Safety external affairs manager. "Contaminants that are typically treated for at Intel are corrosives (acids and caustics) which change the pH of the water; copper and fluoride which are used in the chip fabrication process; and total suspended solids (excessive residual silicon particles that are smaller than a grain of salt that come off the wafer as they are thinned or cut)," he tells me. "In New Mexico," says Niekerk, "we put water back into the river and in Israel we supply water for irrigation. Since 1995, Intel has spent over $70 million on capital costs alone for water conservation systems."

Semiconductor fabs like the one in Chandler exist all over the world. There are fabs in dry countries like Israel, wet countries like Ireland, tropical ones like Malaysia, the Philippines, and Costa Rica, crowded ones like Japan and Taiwan, or rapidly developing ones like China, where industrial growth and construction of all kinds are sucking up water at an unarguably unsustainable rate. Because the success of the semiconductor industry's products depends on a lack of variation, production methods will not vary; so, to be meaningful, resource efficiencies will have to adapt to local conditions.

It's clear that great improvements have been made in water conservation since the days before wastewater was being reclaimed and recycled, but semiconductor fabrication still uses a huge amount of water. As world

water supplies grow scarcer, even more will have to be done to conserve—
or to produce less—perhaps both. Contemplating this dilemma as I sat in
an Intel conference room at the end of my Chandler tour, I wondered how
the semiconductor industry and a company like Intel would deal with the
next challenge on the sustainability horizon—that of remaining profitable
and continually creating new technologies when it's no longer ecologically
viable to increase production volumes *and* dramatically increase computing
power of a semiconductor every eighteen months. That, say Len Drago and
Gary Niekerk, "is the $64,000 question."

As I looked at all the shiny state-of-the-art equipment at the Intel campus, I was reminded how much excitement and glamour there is in new
technology, and in touting environmental efficiencies and moves toward
"sustainable business practices," but not much in looking at decades-old
pollution. But that's where one has to look—back to those old spills and
those aging Leaking Underground Storage Tanks—to understand the origins of concern over high tech's environmental and health impacts.

CHAPTER FOUR

High-Tech Manufacture and Human Health

With a sense of what goes into making semiconductors, I started to track down the environmental and health impacts of some of those chemicals. Stories of groundwater pollution by solvents—particularly trichloroethylene (TCE) and trichloroethane (TCA)—and other toxic chemicals used in semiconductor manufacture can be found in many places that the high-tech industry and its suppliers settled in the United States in the 1970s and '80s or earlier. But Silicon Valley is the place to start.

In the 1970s and early '80s, Silicon Valley semiconductor plants stored large quantities of solvents like TCE and TCA in underground storage tanks—later misleadingly but entertainingly dubbed LUST, for "leaking underground storage tank." "In 1981," Ted Smith of the Silicon Valley Toxics Coalition (SVTC), told me in 2004, "the state health department found a highly contaminated well in south San Jose that belonged to the Great Oaks Water Company—Well 13. The well had high levels of TCA and TCE. In figuring this out, they tracked the source back to Fairchild Semiconductor and found a leaking underground storage tank."

When the Fairchild leak was found, says Smith, "the Regional Water Quality Control Board began doing some testing and found a nearby IBM plant had a leaking underground storage tank which created a major toxic plume. The regional board was sent to survey and created an entire industry database of underground chemical storage tanks. If a facility had one, they began underground monitoring for leaks and found that 85 percent had leaks. It was an epidemic."

"A woman named Lorraine Ross who lived in a south San Jose neighborhood whose daughter had been born with a heart defect read an article about Well 13 in the *San Jose Mercury News* and began to wonder if her daughter's birth defect was related to Well 13," recalls Smith as we sit in SVTC's office in downtown San Jose. "Lorraine Ross started talking to her neighbors and found that many of them had similar problems, and she wrote an open letter to the mayor that was published in the paper. This led to some community meetings, and she talked to us* and my wife, who's an attorney dealing with chemical health issues. They [the community group] wanted to file a series of lawsuits for compensation for the people whose children had been harmed and also wanted to pass a law to prevent what had been happening from continuing to happen."

The discovery of this contamination and community concern about the related health impacts galvanized the grassroots organizing that led to the birth of the Silicon Valley Toxics Coalition in 1982. These efforts brought some of the first public attention to the environmental hazards of high tech and the fact that it wasn't the "clean" industry it had been touted to be. Eventually this activism led to a monitoring program and more rigorous standards for underground storage tanks.

▌▌▌▌▌▌

Trichloroethylene (TCE) and trichloroethane (TCA) are volatile organic, chlorinated compounds.† Neither occurs naturally in the environment. For years TCE has been one of the most commonly used industrial solvents. Its use peaked in the United States in 1970 when it

---

* Smith was then working as a lawyer and had become involved with community, occupational health and safety issues, and defending workers' rights cases.

† Volatile organic compounds are not very soluble in water and have high vapor pressure, meaning they turn to vapor at relatively low temperatures.

began to decline after publication of information about its toxicity that led to its regulation by the EPA in the 1980s.[1] But when TCA—which was sometimes substituted for TCE—and some other volatile organic compounds were discovered to be ozone-depleting chemicals and were banned under 1990 amendments to the Clean Air Act, TCE use began to increase again, despite knowledge of its toxicity and its tendency to linger in the environment.[2]

The U.S. Environmental Protection Agency lists both TCE and TCA as toxic to the nervous, respiratory, endocrine, and reproductive systems, as well as to kidney and liver function. Public health agencies also now consider TCE as a probable carcinogen. Some studies have indicated a link between exposure to TCE-contaminated well water and children born with heart defects.[3] What makes TCE so persistent is that it's heavier than water and not very soluble, so once it enters groundwater it tends to stay there for a long time. Because of its volatility, TCE often passes from groundwater to soil and from soil back to groundwater. TCE also tends to create plumes—pockets or pools of the solvent—that travel within underground water sources. And although it was thought for years that TCE would be contained by soil, it's since been discovered that TCE vapor can pass upward from soil into the air and can travel indoors when contaminated soil comes into contact with cracks in basements, cellars, and foundations.[4]

By the mid-1980s, dozens of leaks and spills of TCE and TCA had been found, not just in south San Jose but also throughout Santa Clara County (as they have been in many communities where industry used TCE). The fast-growing new industry that, to consumers, seemed to create out of thin air machinery both phenomenally powerful and small enough to sit on the head of a pin, had also created an extensive patchwork of groundwater and soil contamination.

Thanks to those leaky chemical storage tanks, Santa Clara County has the greatest concentration of Superfund sites of any county in the country. Over 80 percent of this toxic pollution comes from the high-tech industry,[5] primarily from leaks and spills of volatile organic compounds. Among the other pollutants at these sites are copper, Freon, and lead, but chlorinated solvents predominate. These leaks and spills have affected water supplies that serve hundreds of thousands of Californians, as well as the water quality of San Francisco Bay. Many of these Superfund sites are in densely populated residential areas, some with homes only two hundred feet away.

A Superfund site is, in the EPA's words, "any land in the United States that has been contaminated by hazardous waste and identified by the Environmental Protection Agency as a candidate for cleanup because it poses risk to human health and/or the environment."[6] Even more technically, Superfund is the commonly known name for the Comprehensive Environmental Response, Compensation and Liability Act (CERCLA), enacted in 1980 partly in response to the hazardous industrial pollution of Love Canal in Niagara Falls, New York. CERCLA gives the federal government the authority to respond to releases (or threatened releases) of substances that may endanger public health or the environment. The law also created a federal fund—the Superfund—supported by taxes on the chemical and petroleum industries. The fund pays for cleanups of abandoned or uncontrolled releases of hazardous wastes at qualifying sites and for cleaning abandoned sites when the potentially responsible parties can't be found or fail to act. The high-tech industry Superfund sites are, for the most part, not abandoned sites, nor are they sites where the parties responsible have failed to participate in cleanup activities (although many of the properties have changed owners over the years). Rather the sites involve such extensive and potentially hazardous pollution that Superfund participation was deemed necessary.

Nearly every large high-tech electronics and semiconductor manufacturer that began operations in the 1970s or earlier—Advanced Micro Devices (AMD), Fairchild Semiconductor, GTE, Hewlett-Packard, Honeywell, IBM, Intel, Motorola, National Semiconductor, NEC, Raytheon, Siemens, and TRW Microwave among them—has a Superfund site in its history. And there are numerous sites not on the Superfund list where toxic pollution caused by chemicals used in high-tech electronics manufacturing has also created persistent environmental and health problems. Such sites exist in other communities in Silicon Valley, in Arizona, New Mexico, New York, Texas, and elsewhere around the country and the world. Meanwhile, many communities where chemicals were spilled and leaked in the early days of high-tech manufacturing still suffer from the effects of those toxics, as do a number of former industry employees who worked closely with those chemicals.

||| ||| |

To get a sense of how many people live within walking or drinking water distance of just California's high-tech industry Superfund sites, consider these descriptions from the EPA's documentation of sites discovered in the mid-1990s and earlier:

- "There was contamination from the site that had the potential to reach the deep groundwater that supplies municipal wells within three miles of the site. These wells provide drinking water to approximately 200,000 people in Santa Clara, Sunnyvale and Mountain View."[7] [AMD manufactured semiconductors and microprocessors at this site.]
- "More than 188,000 people live within 3 miles of the site and use groundwater from municipal wells as a source of drinking water."[8] [CTS Printex manufactured printed circuit boards here.]
- "Approximately 300,000 people obtain drinking water from public wells located within 3 miles of the site. The groundwater is used for the municipal water supplies of the cities of Cupertino, Santa Clara and Sunnyvale. Five municipal water supply wells are located within 1 mile of the site. The distance from the site to the City of Santa Clara Well #15, the closest municipal well, is 1,300 feet."[9] [Intersil / Siemens Components, made semiconductors here.]
- "Approximately 189,000 people obtain drinking water from public and private wells within three miles of the site. Groundwater within 1 mile of the site is used for private and municipal purposes. More than 200 private drinking water wells were drilled into the contaminated plume; most of these wells have been closed."[10] [Spectra-Physics made electronic equipment here, and the site adjoins a Teledyne Semiconductor site that has contributed to this pollution.]
- "Contaminants from the site have the potential to migrate to deep drinking water aquifers. Municipal wells for the cities of Santa Clara and Mountain View tap a deep aquifer that serves approximately 300,000 people. The deep aquifer, used as a drinking water source, is not contaminated; however, the shallow

aquifer is contaminated. There are some agricultural wells that could act as conduits for contaminant migration between the shallow contaminated aquifer and the deep aquifer."[11] [TRW Microwave made semiconductors here.]

One of the sites discovered in the early 1980s to be contaminated by an underground leak of TCE was in Mountain View, a Santa Clara County community northwest of San Jose. Since the early days of Silicon Valley, Mountain View has been home to numerous electronics and semiconductor fabrication plants. Fairchild Semiconductor, GTE, Intel, Mitsubishi, NEC, Raytheon, and Siemens are some of the companies that have had manufacturing operations here. NASA's research park and Moffett Field are directly north on the far side of Highway 101. Many of Mountain View's early high-tech businesses are now gone. Internet companies, including Netscape, moved in to take their place, but in spring 2004 when I visit, well after the dot-com bubble burst, many of these properties are empty, and "For Lease" signs litter their shiny tinted-glass windows.

Curious to see what a high-tech Superfund site looks like, I took myself on a tour of Mountain View's Middlefield-Ellis-Whitman site, named after the roads that border the center of the eight-square-mile site. Somehow I half-expected to see a sign that might read: "Attention: You are standing on a Superfund site. The soil beneath your feet is contaminated with TCE, a toxic volatile organic compound recognized by the EPA as a human carcinogen. Were you to dig down to groundwater, you would find that water contaminated by TCE which, in vapor form, is now rising into nearby homes."

Of course there is no such sign. Just a suburban neighborhood of office buildings, single-family homes, and small strip malls with fast-food restaurants, a coffee shop, a gym, a dry cleaner, gas station, and convenience store. I stopped in the coffee shop to get my bearings and wondered briefly, thinking back to a scene in the movie *Erin Brockovich*, if I should worry about the water in my coffee. But no, the residents of Mountain View no longer drink water from the aquifer directly beneath them. Their drinking water now comes from over 175 miles away, from the Sierra Nevada by way of Hetch Hetchy and from other off-site sources. Although the local groundwater is not currently used for munic-ipal drinking water, the City of Mountain View points out it may be needed in the future.[12]

"The Middlefield-Ellis-Whisman area, to me it's the image of the phoenix rising," says Ted Smith when I ask him about the site. "It was the home of the semiconductor industry that engineered and drove the indus-try. They've pretty much all packed up and moved offshore, and it's now home to the Internet."

High-tech electronics companies began operations in this neighborhood in the 1960s and '70s. Contamination here, primarily by TCE, was so exten-sive that beginning in the early to mid-1980s several Mountain View sites were added to the EPA's National Priorities List of locations that qualify for the Superfund program. Some twenty years after discovery of the pollution, cleanup work at most of the Santa Clara County high-tech Superfund sites is far from finished. And it's likely to continue for many years to come. "At one point EPA told us that it would take perhaps three hundred years to get the groundwater back to drinking-water standards," says Smith. "That's longer than this country has been in existence."

Map in hand, I park and wander the properties once occupied by Fairchild, Raytheon, and NEC. The only visible signs of the contamina-tion are some capped wells and a number of gray metal constructions that look a bit like austere public art offerings. These structures are aeration vents, designed to extract the TCE vapor rising from the soil below. A woman is walking her dog in the large empty parking lot where some of these wells and vents are located. I watch them follow the dog's ball into a vacant lot of weeds and wonder if the woman knows what chemicals are percolating beneath their feet.

In the Mountain View Public Library, I read through some of the twenty years' worth of EPA documents about the Middlefield-Ellis-Whitman site. These papers fill huge ring binders and occupy nearly an entire shelf. There are technical reports and related correspondence about water quality, detailed monitoring assessments of the contamination plume, charts re-cording volumes of contaminated water pumped out of the aquifer, of con-taminated soil removal, and details of vapor treatment.

I learn that, in addition to TCE, the contaminants at this site include vinyl chloride, several dichloroethane compounds, TCA, ethyl benzene, xylene, and Freon—all substances toxic to human health and the environment. By 1986 over a hundred monitoring wells had been installed at the site. In 1988 the EPA sealed two abandoned agricultural wells at the site that were carrying the contamination to deeper aquifers, and in 1989 the agency began working on preventing further contamination of the deep aquifer. By the time I'm reading these papers, in March 2004, millions of gallons of contaminated groundwater have been treated and large quantities of contaminated soil have been removed from the site. What's called soil-vapor extraction—work to remove TCE vapor from the soil so that it doesn't enter adjacent buildings—began in 1995, but it will be many years, perhaps decades, before the groundwater will be safe to drink.

Later in 2004 I speak to Alana Lee, the EPA's project manager for the Middlefield- Ellis-Whitman site. She explains that one of the reasons the solvent contamination here is so pervasive is that the water table level has changed over the years due to industrial and municipal development, prior water extraction, and drought cycles. "The contaminated water here moves about one hundred feet a year without pumping," she tells me. A fairly small amount of TCE can thus contaminate a large amount of water. The water also moves vertically, so that as the water table falls and rises, TCE comes in contact with more soil. As the water table is recharged, more water becomes contaminated. And while the contaminated soil removal phase of the cleanup is considered over, groundwater continues to be treated. "Groundwater is the source, and soil is the pathway," comments Lee.

What's now being monitored are levels of TCE vapor rising out of the soil and passing into buildings that sit above and near the contaminated groundwater plume. The EPA does not yet know what the correlation will be between the location of the plume and vapor intrusion into homes and other buildings.[13] Consequently, the City of Mountain View has expressed concern that soil may need to be tested further and that more homes and other buildings may need to be monitored and tested as the plume migrates.[14] "We are viewing this pathway as evolving," says Lee. But the soil-vapor-air pathway "was not looked at back in 1988 and 1989," she says.

"We've only begun looking at it in the past few years." This makes the cleanup timeline difficult to project. "We've been saying decades or over a hundred years," Lee tells me.

‖‖ ‖‖ ‖

California and EPA standards now suggest that any more than five micrograms per liter—or parts per billion—of TCE in drinking water is unsafe. Between 1994 and 1998, TCE levels in groundwater at the Whisman site were found to be between thirty and one hundred parts per billion. The U.S. Occupational Safety and Health Administration limit for TCE airborne exposure in the workplace is one hundred parts per cubic meter—or parts per million—over the course of an eight-hour day, forty-hour workweek. However, some states recommend that air levels of TCE not exceed one part per billion—a factor of a thousand less than what OSHA has stipulated.

You can smell TCE in the air when there are about twenty-five parts per million present. But even if you can't smell it, there may be enough TCE present to do damage if exposure lasts for an extended period. Unfortunately, there are no U.S. federal standards for how much TCE can be present in the air in private homes before it causes harm. State standards, where they have been set, vary widely.

Now, says Lee, it's been discovered that "TCE is potentially more harmful than previously thought, especially to sensitive populations—including pregnant women and children—who are exposed to TCE and its breakdown products." Even back in 1987, a study presented at the American Heart Association's annual meeting showed that children whose parents lived or worked near drinking water contaminated with TCE were over twice as likely as others to be born with heart defects. The study's principle author, Dr. Stanley Goldberg, cautioned that the findings didn't prove that TCE caused the heart defects, only that there was a coincidence warranting further investigation. Yet animal studies done concurrently did show a direct connection between TCE and heart defects. Questions were also raised at the time about the possible carcinogenic effects of TCE, but they were described by a University of California scientist as "controversial."[15]

Although it has become generally accepted that TCE poses serious human health hazards, just how hazardous is still controversial. In 2001 the EPA issued a draft assessment of the toxicity of TCE—which for close to twenty years has been one of the United States' most prevalent ground-water contaminants. This new study found that TCE is likely to be carcinogenic to humans and that TCE may be anywhere from five to sixty-five times more toxic than previously thought.[16]

The 2001 study recommended new guidelines for safe levels of TCE that could reduce the current drinking-water standard from five parts per billion to less than one part per billion. It also recommended that inhalation standards be revised from about one microgram per cubic meter to about 0.016 micrograms per cubic meter. But these findings were greeted with criticism, particularly from the solvents industry, and the EPA has referred the report to its science advisory board for review.

As of 2005 this assessment had not yet been finalized, and the board's findings may not be ready until 2006. Consequently, different parts of the country dealing with TCE contamination of water and indoor air are using widely varying safety standards—a situation that residents of affected communities find troubling. This is especially disconcerting because a level deemed safe one day was changed dramatically downward on another and because residents of a number of these communities may have been exposed to TCE by way of indoor air over long periods of time. This concern is compounded by the fact that many people who live in communities affected by TCE contamination may also be exposed to hazardous chemicals on the job.

BEYOND THE BUNNY SUIT:
TOXICS IN THE WORKPLACE

One person who has been working to reduce the levels of toxics present in the workplace and to protect those who may have been harmed by them is Amanda Hawes, an attorney with the San Jose–based firm of Alexander, Hawes and Audet. In 2004 Hawes received news media attention for her representation of former IBM employees who sued the company for cancers they believe were caused by chemicals they were exposed to during their work in clean rooms at IBM's San Jose semiconductor plant. The case was brought on behalf of Alida Hernandez and Jim Moore, who charged that IBM had exposed them to TCE, Freon, benzene, and other chemicals at the plant between the 1960s and the '80s. Proving that IBM had knowingly allowed Hernandez and Moore to be poisoned by toxic chemicals used in the course of their work, and that these former employees' cancers were caused directly by these chemicals, was not easy, especially given the standards for proof in such cases. In June 2004 the suit was settled in IBM's favor. Over two hundred comparable cases were filed against IBM, however, and as of this writing over one hundred remain to be heard.

"In a claim by a worker in the courtroom, it's not a level playing field," says Hawes in September 2004, speaking on the phone from her office in San Jose. "It's slow, it's expensive, but I can't sit back and say this can't be done. I fight for keeps and I do it for clients who didn't get a chance."

Hawes's advocacy on behalf of those working in Silicon Valley began in the 1970s, before bumper crops of semiconductors and circuit boards had completely eclipsed those of the region's orchards. "My passion and principle concern has always been for workers. My focus is to protect workers who get the brunt of it," Hawes tells me. In the mid-1970s Hawes began working on behalf of cannery workers, on discrimination cases, on issues of pay equity for women workers, and on health issues, including those that involved the impact of chemicals. "I thought we could do something to improve working conditions for everyone." Unfortunately, she says, "It's never been the case that worker health and safety is a priority."

By 1975 and '76, the Valley canneries had begun closing, and former workers there began taking jobs in high tech. "This looked like a tremendous opportunity," says Hawes, for these workers to move into a "clean industry." But she and her colleagues soon found the industry's use of chemicals to be disturbing. To address these issues, Hawes and others began a group called Electronics Committee for Occupational Safety and Health, which by 1978 had evolved into the Santa Clara Center for Occupational Health and Safety (SCCOSH). The SCCOSH offices are located, not coincidentally, next door to those of the Silicon Valley Toxics Coalition to whose founder, Ted Smith, Hawes has been married for thirty years. Their work has often dovetailed, and whether working independently or in tandem, over the years they have become perhaps the most effective advocates for improving environmental practices throughout the high-tech industry.

As Hawes wrote on the occasion of SCCOSH's twentieth anniversary in 1998:

> In the mid 70's job health and safety in semiconductor fabs usually meant protection from acid burns, knowing where the eye wash station was, and storing chemical bottles in locked cabinets. Compared to hazards of traditional fruit processing in our Valley of Heart's Delight—repetitive motion injuries, finger lacerations, heat stress, and slips and falls—conditions in Silicon Valley's "clean industry" looked good, especially to workers laid off after years of back-breaking seasonal work in the canneries. At the same time, a small group of people was meeting to discuss concerns over the chemical-handling aspects of this industry and what might be done to raise these issues publicly and before workers started paying with their health for Silicon Valley's enormous success . . . Our efforts were met with politeness and a fair amount of skepticism: how could the "clean industry" be hazardous to your health?[17]

As of late 2004 there were about 255,000 people employed in the semiconductor industry in the United States—a number that has been reduced by 30,000 or more over the past few years as chip fabs have become increasingly mechanized and production overseas has burgeoned.[18] It has proven hard to get a firm number of how many people are employed in the semiconductor industry worldwide, but one estimate puts the number at about one million or more.‡[19] "This is a forgotten group of workers," Dr. Joseph LaDou, director of the International Center for Occupational

---

‡ When I asked SIA spokesperson Doug Andrey about this number, he said it was probably correct in order of magnitude but that he didn't know how one would even go about finding such a figure and cautioned about the dangers of extrapolating. The SIA figure is for U.S. plants; it doesn't include employees of U.S. companies' non-U.S. facilities.

Medicine at the University of California at San Francisco, told *Mother Jones* magazine in 2002.[20]

That the public face of the industry rarely includes the people who work the machines that make the physical guts and brain of high-tech electronics isn't entirely surprising. What we've been presented with is the tidy finished exterior of this equipment, its power to manipulate information, and the high-speed communication it enables. We haven't been told the story of the person who worked a twelve-hour shift in Taiwan and developed a urinary tract infection because she didn't want to take breaks for drinks of water or the bathroom[21]—or the worker in upstate New York who developed nosebleeds while working with photoresist chemicals.[22]

Among the first issues SCCOSH tackled were the health impacts of TCE. Although the group's campaign to achieve a legal ban on TCE was not successful, the group did succeed in helping to move the government— state and federal—to set limits for human exposure to TCE in the workplace. And as a result of increased knowledge—and publicity—about TCE's hazards, the major semiconductor manufacturers have largely discontinued its use. But the chemical is still permitted. "If we could get it out of the workplace, we would get it out of the waste stream," says Hawes of TCE and similar chemicals. For it's in the waste stream that TCE's potentially carcinogenic effects linger for years, with the potential to affect many more people than those who might be exposed to it at work. "There's no precautionary principle at work here, and that's very unfortunate," says Hawes. "If something's a carcinogen, it shouldn't be there in the first place."

Another group of chemicals whose use in semiconductor production concerned SCCOSH were glycol ethers, a class of solvents used in the photoresist step of the chip-etching process. There are two kinds of glycol ethers; the ones that are toxic to humans are ethylene-based. These are used in chip fabrication. According to the California Department of Health Services, "Overexposure to glycol ethers can cause anemia . . . intoxication similar to the effects of alcohol, and irritation of the eyes, nose or skin. In laboratory animals, low-level exposure to certain glycol ethers can cause birth defects and can damage a male's sperm and testicles . . . Based on the animal tests and on studies of workers, you should treat certain glycol ethers as hazards to your reproductive health."[23]

Beginning in the early 1980s, Hawes tells me, women who worked with these chemicals in semiconductor fabrication had been reporting what she describes as "a very significant pattern of pregnancy loss and subfertility." In 1981, following a study by the National Institute of Occupational Safety and Health, SCCOSH published a report that showed exposure to glycol ethers causes reproductive health damage in animals.

A year later the California Department of Health Services issued its health warning on glycol ethers. At the same time the Semiconductor Industry Association (SIA), which includes about 90 percent of all U.S. semiconductor manufacturers, made its members aware of these findings. Between 1982 and 1989—while glycol ethers continued to be used in chip fabrication—additional studies documenting the reproductive system toxicity of glycol ethers were published, as were at least two epidemiological studies showing the connection between miscarriages and exposure to glycol ethers.[24] But before learning more about glycol ethers, I wanted to know how it was possible for semiconductor plant workers to be exposed to toxic chemicals if they worked in what are called clean rooms and wore "bunny suits."

HOW CLEAN IS A CLEAN ROOM?

Clean rooms are the hub of semiconductor fabrication. This is where a significant portion of wafer fabrication takes place, where the transistors that form a chip's circuitry are etched onto silicon wafers, the dopants are introduced, metallic elements added, and the circuitry tested. Nearly all of the processes that go into turning a silicon or other semiconductive wafer into a microprocessor take place in a clean room.

To understand what a clean room is, you can take a virtual tour of Intel's chip fabrication plant in Ireland. "Before we begin our journey through the Fab," says our virtual tour guide, "we must first put on a special 'space-age' suit . . . We need this very clean environment because the transistors we are building on the silicon are 500 times smaller than the width of a human hair. Any speck of dust or dirt falling on the wafers would ruin the microchips."[25]

A chip can be ruined not only by dust or dirt, but also by a single hair, flake of skin, or anything else a human might possibly shed. So clean rooms are built out of materials that don't release any particles, with air filters and circulation systems designed to keep out the smallest particles of dust and debris.[26] Those who work in a clean room wear head-to-toe coverings known as "bunny suits"—as well as gloves and shoe coverings—all changed after a single use. Some of these suits even have individual air-filtering systems. While it looks as if bunny suits were designed as a safety measure for those who wear them, clean rooms and their accessories were designed to protect chips, not to protect those who are making them. And free of specks of dust and dirt doesn't necessarily mean free of chemical vapors.

Critics of clean-room conditions point out—as they did in the California IBM workers' suit—that the air in these rooms is typically recirculated. This means that workers may be exposed to recirculating toxins, which have the potential to cause damage from repeated exposure over time, even if initial exposures are at or below legally permitted safety levels. Clean rooms have specially designed ventilation systems positioned over workstations where steps in chip fabrication take place—a kind of high-tech version of the hood vent you might have over your home stove. But these hoods haven't always been 100 percent effective at removing airborne chemicals, and the bunny suits and clean rooms seen today didn't come into use until the 1980s, in some places not until the early 1990s. This means that semiconductor fabrication workers who began their careers in the 1960s and '70s may have had considerable exposure to chemicals used in chip production.

Over the years, work in U.S. clean rooms has become increasingly mechanized, so that fewer workers are involved. When I toured an Intel fab in Chandler, Arizona, in July 2004, I saw far more machines than people, and the people I chanced to see were checking computer screens rather than manipulating chip-making machinery. From behind sealed windows and doors, looking into the 160,000-square-foot clean room areas (some clean rooms are as large as 200,000 square feet), I could see the enclosed stations where some of the wafer-etching steps take place, the complex machinery that lays in the metal ions, and the mechanized carriers that ferry chips from one part of the fab to another. The few workers I observed were indeed wearing bunny suits and protective face gear, and there were huge cardboard bins of used latex gloves awaiting recycling.

But well into the 1980s, chip manufacturing was very much a hands-on job performed in ways that brought workers into close contact with ethylene-based glycol ethers and other chemicals. In the early days of semiconductor production, for example, photoresist was applied and removed manually. Workers would dip a circuit panel or wafer in an open container of photoresist solution and then sandwich it between panels of glass. These glass panels were used over and over again, and between each use they were hand-wiped with solvent-soaked rags. To wipe a panel meant plunging a rag in a can full of solvent or dipping the panel directly into a large tank of TCE. The etched wafers were then hung up to dry on a traveling circular rack. Into the 1970s this work was done without wearing special protective clothing, according to Rick White, a twenty-eight-year veteran of IBM's microelectronics plant in Endicott, New York. No masks were worn, he told me in 2004, and workers were given gloves only if they insisted on them.

White, a second-generation IBM employee, worked for years in the Endicott campus's Building 18, where most of the chemical-intensive semiconductor manufacturing took place. "Building 18 was a horrific place to be," says White. "When you walked in, there was a smell that you can't get out of your nose. If I had known then what I know now," he says shaking his head. White says it wasn't until the late 1970s that safety courses began at the Endicott plant.

"There was a rust-brown syrupy liquid we used to strip the photoresist off the panels," White tells me. The dipped panels would be hanging up in the air on circulating racks for an hour to an hour and a half, he says. There were exhaust fans, but they'd shut down maybe once or twice a day, and an alarm would go off.

One of White's jobs was to wipe the glass panels then used in the photoetching process. Among the substances White worked with were TCE and methyl chloroform (another name for trichloroethane or TCA). "My nose started getting dry and I started to get nosebleeds every night I worked in that room," says White. So he went to the on-site doctor. I told him, "I think it's something I work with," White recalls. The doctor didn't agree; he told White his nose was dry because he didn't drink enough water. White was told to swab his nostrils with a salt, sugar, and water solution three times a day. His own doctor suggested instead that he ask for a transfer to another area of the plant. Eventually, White was able to get an office job—rather than doing hands-on manufacturing—as a support person in the engineering department.

Wanda Hudak, who has been a Broome County legislator and worked as a nurse at IBM's Endicott plant from 1969 to 1979, confirms White's recollections. No-nonsense, good-humored, and often characterized in the local press as "an outspoken Republican," Hudak says that in those days there might have been from three thousand to six thousand workers on a night shift and that she might see up to ninety cases a night. Most were minor, but occasionally there were accidents involving chemicals. In the case of accidents, she tells me, the first thing Hudak would do was to "get the guy into the shower and look the chemical up in a reference book."

"There were guys who were working with the chemicals for twenty years who'd come to me and say, Mrs. Hudak, I've got this rash and it doesn't seem to be going away," says Hudak. She was asked to keep track of which employees were coming in for medical attention as part of an effort to cut down on what she called "malingers." But "I refused to give them names," she says. "There wasn't a lot of understanding on the part of the company."

When reports of miscarriages and related problems among women who worked in clean rooms began to accumulate through the 1980s, chip makers allowed women who were pregnant to opt for work in other divisions. Some later advised against pregnant women working in clean rooms. These recommendations came about because, in Amanda Hawes's words, in 1992 "two shoes dropped." First, a decade after the California Department of Health Services first issued a warning about the health hazards of ethylene-based glycol ethers, the Semiconductor Industry Association's own study—undertaken in 1989 by the University of California at Davis—was released. That study, which surveyed over fifteen thousand workers at fifteen different companies, found that women working in semiconductor fabrication plants who were exposed to glycol ethers and other chip-etching chemicals did indeed have higher rates of miscarriage than women working elsewhere in these plants.

Second, a similar study was also released in 1992, this one commissioned by IBM and conducted by Johns Hopkins University. Its findings mirrored what the SIA epidemiological studies had found. Similar reports of miscarriages and of reproductive system and other health problems were also being made by women who had worked at the National Semiconductor fabrication plant in Greenock, Scotland, and, later, by women working in semiconductor plants in Taiwan and elsewhere around the world. As a result of such studies, and continued agitation on the part of high-tech industry watchdogs and worker advocacy groups—chief among them the Silicon Valley Toxics Coalition and SCCOSH—the semiconductor industry began phasing out the use of certain ethylene-based glycol ethers in the mid-1990s.

▌▌▐▌▌ ▌

Yet despite the health hazards associated with these glycol ethers, the EPA's year 2000 "hazard summary" for these compounds—the most recent available as of this writing—says, "No information is available on the reproductive, developmental, or carcinogenic effects of the glycol ethers on humans."[27] The summary did note that, despite this lack of information, "a possible effect on sperm quality and testes size in workers exposed to glycol ethers has been reported" and that "animal studies have reported testicular damage, reduced fertility, maternal toxicity, early embryonic death, birth defects, and delayed development from inhalation and oral exposure to the glycol ethers."[28]

Despite these hazards, at the end of December 2003 OSHA noted in the *Federal Register* that the Department of Labor had decided against issuing a final safety standard for this class of glycol ethers. Evidence of adverse reproductive and developmental health effects notwithstanding, OSHA, says the *Federal Register*, "has decided to terminate the rulemaking because production, use and exposure to these glycol ethers has ceased or is virtually limited to closed system production where there is little opportunity for employee exposure. Exposure levels in those operations already are at or below the proposed PELs [permissible exposure limits]. In addition, use of these glycol ethers has largely been replaced by less-toxic substitutes."[29]

While this is good news for those who work for U.S. chip manufacturers, this also means that, despite the health hazards, there has been no federal regulation barring glycol ethers, nor has the federal government issued any definitive safety standards for human exposure to these toxic chemicals. In the European Union, on the other hand, these glycol ethers have been banned from use in consumer products since 1994.[30]

But the story does not end there.

Among the cases that have been filed against IBM by former workers at its New York and Vermont plants are a number that involve birth defects these employees claim were caused by glycol ethers and other chemicals used in chip manufacture. The details of these birth defects are horrific, heartbreaking, and almost too grim to detail.

As Bob Herbert wrote in a 2003 *New York Times* Op-Ed piece, taking up the cause of these workers' children, "There is a long list of young people and children who have suffered tragic birth defects—spina bifida, missing or deformed limbs, a missing kidney, a missing vagina, blindness—whose parents (in some cases both parents) worked for IBM and are now suing."[31]

Two of these cases filed by former workers at IBM's East Fishkill, New York, plant have been settled out of court, including one that involves a teenager named Zachary Ruffing who was born blind and severely deformed. The other, brought by Candace Curtis, a twenty-three-year-old woman born with brain damage and with no kneecaps whose mother worked at the plant was settled in March 2004, the day jury selection was scheduled to begin. At the end of 2004 there were still about one hundred similar cases involving cancers and birth defects pending in New York courts against IBM.[32]

The industry itself, not surprisingly, asserts that their manufacturing processes are not to blame. An IBM spokesman quoted in the *Wall Street Journal* said of the Curtis case, "We believe the evidence and facts will prove the workplace didn't cause the physical problems," and that chemicals the workers at their plants were exposed to "were at or below the levels established by health and environmental regulators."[33]

"We need to address the huge discrepancy between what's permitted and what's safe," Amanda Hawes counters. "It's critical that we try to eliminate this difference." It's as if you said people were driving safely when they "are driving at 200 miles per hour when the speed limit is 2000 miles per hour," she adds.

‖‖ ‖ ‖

In March 2004 the Semiconductor Industry Association announced that Bureau of Labor Statistics findings for 2002 showed that the incidence of work-related illness and injury for the industry was only 1.9 per 100 full-time workers, better than 95 percent of other comparable manufacturing industries.[34] Environmental health and safety data from chip-making giants Intel, IBM, and AMD for 1999 through 2003 show even smaller rates of recorded occupational illness and injury—for some years, rates are less than one case per hundred workers.

Curious how such statistics could exist concurrently with studies that detail high rates of illness and with anecdotal reports from high-tech workers themselves, I checked in with the Bureau of Labor Statistics. First, it helps to know that the agency's figures for occupational injuries and illness are compiled from reporting that OSHA requires of businesses over a certain size. What these businesses report are immediate incidents of injury or illness that occur while a worker is on the job. This reporting wasn't designed to track health problems that might emerge after a worker has left the job or retired, so it doesn't reflect any incidence of delayed illness. It may be years before cancer, reproductive problems, birth defects, or any other health problems associated with chemical exposure become apparent, and it's notoriously difficult to pinpoint a single trigger or source for these health problems, especially using current U.S. methods of assessing risk when environmental factors may be involved.

The SIA's cancer risk study released in 2001, for example, found no definitive connection between working in chip fabrication and increased cancer risk—but the study was based on a review of available scientific literature, and no studies were available specifically identifying semiconductor workers.§ Offering little help for those hoping for a definitive answer, the report concluded,

1. There is no affirmative evidence at the present time to support the contention that workplace exposures to chemicals or other hazards in wafer fabrication, now or historically, measurably increase the risk for cancer in general, or for any particular form or type of cancer.

2. Conversely, there is insufficient evidence at the present time to conclude that exposures to chemicals and other hazards in wafer fabrication have not or could not result in measurably increased risk of one or more cancer types.[35]

Other studies analyzing U.S. semiconductor manufacturers' records have reached a different conclusion about the industry's cancer risk. One such study, done by Boston University epidemiologist Richard Clapp and private consultant Rebecca Johnson, has caused an uproar in the scientific publishing company. Clapp and Johnson analyzed IBM's health records and concluded that workers at IBM's San Jose plant were two to six times more likely to develop certain cancers than is the national average. When the journal *Clinics in Occupational and Environmental Medicine* declined to publish the study after it had been accepted by the issue's guest editor, twelve other contributors threatened to withdraw their work in protest.[36] IBM attorneys had asked that Clapp and Johnson's report not be made public while the lawsuit of Alida Hernandez and James Moore was in progress, and the study's findings were not allowed to be used in the trial. As of summer 2005 the report had not yet been published.

Legal maneuvers and the politics of peer-reviewed journals aside, what the semiconductor industry's health studies (including its 1999 cancer risk studies[37]) didn't do was investigate what effect exposure to gly-

---

§ The health study announced by the SIA in March 2004 may help to partially fill in this gap. It will be "a retrospective epidemiological study to investigate whether or not wafer fabrication workers in the U.S. chip industry have experienced higher rates of cancer than non-fabrication workers" (SIA press release, March 2004).

col ethers and other semiconductor fabrication chemicals may have had on the babies born to women who worked in these plants. Nor did the studies look at another aspect of potentially considerable importance: what the effects of exposure to multiple toxic chemicals over periods of time may be.

"In the workplace, people are routinely exposed to many chemicals at the same time. The industry studies that are done are for single chemicals. I don't think enough has been done to own up to the hazards of workers in mixed chemical environments," says Amanda Hawes. And relatively little research has been done on the health and developmental effects of exposure to multiple chemical toxins.

High-tech manufacturing and the proliferation of Information Age electronics have released into the environment numerous synthetic chemicals. Many of these compounds are persistent and hazardous to human health. Some have adverse impacts on the endocrine and nervous systems. Some have been shown—even at very small doses—to affect embryonic cell development, and some have been linked to behavioral, developmental, and neurological disorders, as well as to autoimmune and other chronic diseases. Simply by being alive in the world today, most of us will likely be exposed to one or more such compounds on a daily basis. What's of particular concern is that children born in the past twenty years have been exposed to these chemicals at particularly vulnerable stages of their development.[38] But most of the research done thus far has focused only on single chemicals.

Interested to know what a working scientist might have to say on the subject, I spoke to Carol Reinisch, a scientist at the Woods Hole Marine Biological Laboratory. Reinisch studies the effects of multiple pollutants—or what she calls a "chemical cocktail"—on developing cells. She is an expert in PCB-induced neurotoxicity and coauthor of an EPA-funded study of the effects of several pollutants—bromoform, chloroform, and tetrachloroethylene—on nerve-cell development in clam embryos. This paper, write Reinsich and her colleagues, "is one of the first reports demonstrating that a mixture of environmental pollutants can act synergistically to alter a critically important biochemical pathway in the developing embryo."[39]

"Looking at single chemicals doesn't make a lot of sense to me because we are exposed to multiples," said Reinisch in January 2005. "Frankly," she said, "this science is in its infancy."

In addition to the problems posed by multiple chemical exposures, another factor complicating the effects of workplace and environmental toxics exposure is the fact that people respond to these compounds differently. Body type, personal history, and behavior all influence how a body reacts, creating a complexity that often blurs clear lines of cause and effect in the legal arena.

"Every human being has a remarkable set of factors and conditions that may elevate their risk of developing cancer . . . To look at snapshots here and there and not consider an array of other risk factors is unreasonable and unscientific," said Robert Weber, an attorney representing IBM in the Hernandez and Moore case, in remarks quoted by the Associated Press.[40] He noted that Hernandez is overweight and diabetic and that Moore is a smoker who was exposed to toxic chemicals while working in orchards and a gas station. These personal histories didn't help Hernandez and Moore's case in court, but Weber's statement underscores the need for greater understanding of how different risk factors—congenital and environmental—may contribute to health problems and disease. His comments also show the need for a legal and regulatory system that errs on the side of protecting human health and the environment rather than waiting for multiple incidents of severe illness, injury, or even fatalities to declare a substance unsafe.

With legislation like RoHS and WEEE influencing what happens in high-tech manufacturing worldwide, and with REACH legislation likely to effect chemical producers everywhere, globalization has the potential to play a role for good. "I don't want to overemphasize the good," Ted Smith tells me of the WEEE and RoHS directives, "but taken together, the two directives are having a tremendous impact in harmonizing things upward."

If the multinational nature of high-tech manufacturing and increasing demands for transparency can impose high environmental and safety standards on the industry as a whole, there should be an improvement in working conditions throughout all companies that supply components for European and U.S. manufacturers. Mandatory environmental performance reporting is likely to spread beyond Europe, says a report by Enhesa, a global environmental health and safety consulting firm based in Brussels.[41] The report cites the prevalence of such reporting among U.S. companies and guidelines that have been set forth in Malaysia, Japan, and Australia. However, concerns remain that as high-tech manufacturing spreads, countries not participating in such programs will continue to allow businesses to operate under conditions that threaten worker and environmental health and safety.

Has there been progress, I ask Amanda Hawes? "I haven't expected a whole lot from the government, and the current level of federal support [for this kind of issue] is very discouraging. Nothing's really happened on a federal level since the Carter administration," she says. Her words will come back to me when I visit Endicott, New York, one of many communities around the world where the high-tech industry's past practices continue to haunt its present and create worrisome uncertainty for those who live with this legacy.

HIGH TECH IN THE RUST BELT

Three thousand miles from Silicon Valley, in upstate New York—a landscape more often associated with the Industrial Revolution than with the Information Age—a plume of contaminated groundwater that encompasses over three hundred acres is drifting toward the Susquehanna River. In Endicott, New York, where IBM began manufacturing business machines and their components in the 1930s and started fabricating semiconductors over forty years ago, at least 480 homes run water that contains trichloroethylene.

As it has in Mountain View, the TCE and TCA have permeated Endicott's aquifer and passed from the groundwater into the adjacent soil. The TCE has been in the groundwater since at least 1979, quite likely longer. TCE vapor is now seeping from the soil into the community's homes and businesses. Close to five hundred Endicott properties are being monitored for TCE vapor intrusion. Congressman Maurice Hinchey, whose district includes Endicott, has called the groundwater pollution "a sleeping giant."[42]

On a chilly November Sunday, with the remnants of the previous days' snow clinging to front yard shrubbery, Alan Turnbull, director of a local citizens' group called Resident Action Group of Endicott (R.A.G.E., for short), gives me a tour of the part of town that sits above the contaminated groundwater. Turnbull became an activist in 2002 after his wife Donna, who is not a smoker,** was diagnosed with a relatively unusual form of cancer that affects the throat. When this happened, "I decided I was going to get to the bottom of this. The more questions I asked, the more questions I had, and the more questions other people had," Turnbull told me as we drove the quiet streets of Endicott.

Endicott calls itself the "Birthplace of IBM," which sounds rather grand, but in late 2004 Endicott, with its population of just over thirteen thousand, has the look of a hometown left behind for shinier, more cosmopolitan locales. About an hour's drive south of Syracuse and ten miles west of Binghamton, the Village of Endicott is very much a Rust Belt manufacturing town. It grew up around two companies, the Endicott Johnson Shoe Company and IBM.

"E-J," as the shoe company is known locally, built many of the town's modest two- and three-story wooden and brick homes for its workers and established parks and other community recreational facilities. Endicott Johnson, which had roots dating back to the Civil War, is now history. But at one time it was the largest shoe manufacturer in the world, turning out fifty thousand pairs of shoes a day.

In its heyday in the 1980s, IBM employed well over twelve thousand people in Endicott; at one point IBM's employees in this part of New York State, an area known as the Southern Tier, numbered close to twenty thousand. In 2002 IBM sold its property to a company called Huron Associates, but it continues to operate at its former 150-acre electronics manufacturing site that dominates the middle of town. As of 2005 IBM employed only about fifteen hundred people in Endicott.

The leitmotif of my tour are the flat-topped cylindrical monitoring

---

** In the interests of full disclosure, I must note that Alan Turnbull is a smoker, although during my visit he smoked only on the porch.

wells that are elsewhere in downtown Endicott. About three feet high and maybe a foot or so in diameter and painted a neutral brown, these wells—designed to test the groundwater below for TCE and other solvents—are unobtrusive and might go unnoticed. But there are over three hundred of them, Turnbull tells me, scattered across the more than three-hundred-acre area of the contamination plume. Some are planted in the sidewalk near the old IBM headquarters, a Federal-style brick building with the word "THINK" carved above the white cornices and columns. Some of the monitoring wells stand in front of the offices adjacent to the IBM manufacturing facilities and its parking lot.

Turnbull points out the white ventilator pipes, part of the TCE vapor monitoring and extraction system for homes sitting above the contamination plume. If you didn't know what you were looking at, these pipes, like the monitoring wells, would pass unnoticed or be mistaken for an odd piece of gutter or downspout. Turnbull also shows me where, for a number of months, a trailer housing an office to answer Endicott residents' questions about the groundwater contamination had been set up. The trailer is gone, but the residents' concerns are far from answered.

Endicott residents still have ample reason for worry and confusion. There are questions about the amounts of chemicals spilled, about the health hazards they pose, about exactly how and where community residents may be exposed to them, and about measures being taken to safeguard health and drinking water safety. These questions were slow to surface, but once residents began to worry not only about their health, but also about how the TCE vapor intrusion might affect their property values, people gradually began to speak out.

"The feeling here was that IBM was taking care of it and mostly we forgot about it," says Turnbull of the contamination. "People didn't even know the chemical names."

That was before residents of Endicott—and the surrounding communities that use the Endicott water system—were fully aware that volatile organic compounds, including TCE, were being detected in their drinking water on an ongoing basis. Which also means TCE is in the water they bathe in, the water they brush their teeth with, the water they cook with, wash dishes and clothes in, water lawns and gardens with, and let their children play in on a hot summer day. And it was before hundreds of residents realized they may have been breathing TCE vapor for years, in some cases, decades.

The night before I arrive in Endicott, I go through the newspaper clippings and documents I've gathered. After reading about someone whose son suffered a severe instant allergic reaction after showering in water known to contain TCE, I decide—quite likely in a case of foolish overreaction—to stay in a motel that draws water from a source other than the Endicott aquifer. But this gives me a taste of what Endicott residents are dealing with—a potential poison they can't see or smell, present at concentrations that may be harmful, and whose damage may not emerge for years.

Some of the Endicott groundwater contamination has been traced to the shoe-company site, some to a town dump, and some to various other business and industrial sites (including a dry cleaner), but the largest documented source is the IBM campus on North Street in downtown Endicott. According to publicly available New York State Department of Environmental Conservation (DEC) reports, a spill of volatile organic compounds at IBM's Endicott plant occurred in 1979—a spill of about 4,100 gallons. But in 1980 IBM notified the DEC that the company had in fact found that tens of thousands of gallons of chemicals had leaked or spilled and were pooled beneath Building 18, the heart of the company's Endicott manufacturing operations. The following year IBM began to clean up the contamination by pumping, extracting, and treating the tainted water. Twenty-five years later that pumping is still going on. Thus far about 80,000 gallons of volatile organic compounds have been extracted.

Some Endicott residents question whether or not this contamination actually stems from the single spill reported in 1979 given the quantity of the contaminants that have been extracted and the volume IBM reported finding pooled under Building 18 in 1980. Could that all have come from one spill of 4,100 gallons? And it wasn't until the late 1970s, after the passage of the Resource Conservation and Recovery Act in 1976, that the EPA gained the authority to control hazardous waste.

Correspondence from 1977 and 1978 between IBM's Endicott plant and the DEC, some of which was copied to the EPA, reports on spills of copious amounts of "industrial waste."[43] On February 3, 1977, IBM reported that a break in a water transmission line allowed rinse water—about 40,000 gallons—from the plant to be diverted to the storm sewer for six hours. On May 24, 1977, IBM reported a valve failure that caused a discharge of 45,000 to 50,000 gallons of industrial waste to the storm sewer. Six months later, IBM reported a spill of approximately 450 gallons of methyl chloroform, about half of which flowed into "the industrial waste stream." Another spill of 1.75 million gallons of wastewater containing what IBM described as "an abnormally high discharge of methyl chloroform" occurred on the night of January 26, 1978. About eight weeks later, IBM wrote to the DEC about a discharge of "untreated industrial wastewater" that spilled into the Endicott storm sewer for two hours at the rate of approximately 100 gallons of minute—which would mean about 120,000 gallons. This discharge contained chromium, hexavalent chromium, copper, iron, and cyanide.[44]

These documented spills seem to indicate that Endicott's groundwater contamination is unlikely to have resulted from isolated accidental discharges. This correspondence seems to indicate a chronic incidence of equipment failure and or human error that allowed large amounts of water laced with toxic solvents and other chemicals to enter the local water system. A legacy of sloppy practices over a long period of time is nothing unusual for a site like this, acknowledges the DEC. Still, DEC documents consistently say that the IBM Endicott water contamination began in 1979. On the other hand, when I asked IBM spokesman Tim Martin about this in 2004, he responded, "We have never indicated that one 1979 release of TCE accounts for the entire Endicott situation."

There are also questions about why, in 1986, the Endicott groundwater contamination's DEC classification was changed. Despite the persistent nature of the contamination, and the fact that the DEC says "it typically takes many years (possibly decades) to remove VOCs [volatile organic compounds, like TCE and TCA] from groundwater,"[45] in June 1986 the DEC changed the classification of this site's contamination from what's known as a Class 2 site—one that poses a threat to public health—to a Class 4 site, a case that has been properly closed, presumably cleaned and no longer a threat. That this reclassification took place when state environmental authorities knew that groundwater contamination was extensive and complex "is among the most glaring problems at the site," says Congressman Maurice Hinchey, who has spoken publicly about "shoddy record keeping" and of "a deliberate strategy by polluters to avoid admission of culpability."[46] As is common in such cases of industrial pollution, IBM is cooperating and participating with the cleanup—and has been for the past twenty-five years—but without any admission of wrongdoing.

What's happened in Endicott since the discovery in 2002 that TCE vapors were rising through the soil above the contamination plumes has been fairly byzantine. In July 2003 the DEC said that it didn't plan to reclassify the groundwater contamination as hazardous, prompting an outpouring of frustration and anger from Endicott residents. Yet that same month the New York State Department of Health said it would initiate a study of area residents to determine what, if any, health problems might be connected with exposure to TCE and related chemicals. At the same time, the Village of Endicott decided to build an expensive filtration system for the community's water and accepted a $2.1 million gift from IBM for this system, which has yet to be built.

In February 2004 the DEC announced that cleanup efforts had *not* significantly reduced the amount of toxic chemicals in the groundwater. Therefore, the DEC said it would reclassify the site as Class 2, one that poses a threat to public health and the environment. Why documentation determining the site's original reclassification—from hazardous to closed—is missing has not been determined. The original documents about IBM's cleanup of the site have also gone missing. And there are boxes of documents related to the contamination that have yet to be scrutinized. This, Congressman Hinchey has said, contributes to "mistrust that shrouds this whole affair, leading many people to suspect that innocent people have been exposed to toxic chemicals."[47]

In August 2004 the DEC formally asked IBM to speed up the pace of the cleanup and outlined in a legally binding document what must be done in the subsequent year. But estimates are that it will still take at least ten years to remove up to 80 percent of the contamination. Meanwhile, Endicott residents continue to live with uncertainty about the safety of their air and water.

Throughout the 1980s and nearly all of the 1990s, it was assumed that if TCE was found in groundwater, it would remain there. The possibility of TCE vapor intrusion was not contemplated until it was discovered. "It's only recently, in the last three years, that the issue of indoor air pollution has come to light," said New York State Assemblyman Tom DiNapoli, speaking to a standing-room-only crowd at a public hearing in Endicott on November 15, 2004. "The degree of uncertainty associated with these issues is an issue in itself," said DiNapoli, who called for "a preventative approach" in regulating these chemicals.

"These were commonly used solvents," says IBM spokesman Tim Martin. "Do we wish that science had the kind of understanding of chemicals in the 1940s and '50s as we do now? Of course we do," he says. "IBM has always been considered to be a leader in environmental management, and about 80 percent of what we know [about contamination like this] comes from this project. This is a case where it's not good to be first. But IBM has performed consistently."

Beginning in 2002, prompted by the EPA's investigation of a similar situation involving TCE contamination in Colorado—and about the same time that TCE vapor intrusion began to be detected and monitored in Mountain View, California—the DEC asked IBM to help assess whether TCE vapors were rising from Endicott's groundwater, passing through the soil and up into buildings. In 2003 hazardous levels of TCE vapor were found in about 75 Endicott properties. By the end of 2003 the number of homes and businesses being monitored and tested for TCE vapor had climbed to 480. And more affected properties may well be found as the plume moves and vapor migrates.

"Even at low concentrations these chemical contaminants can lead to chronic health problems," said Carl Johnson, deputy commissioner at the DEC's Office of Air and Waste Management. This comment at the November 2004 hearing confirmed what many of the assembled Endicott residents feared.

"TCE appears as the most pernicious and threatening VOC that we are facing," said Congressman Hinchey at that same public hearing. "The federal government," he continued, "has been negligent in promulgating new standards, and the efforts to revise standards have met with strong resistance by the U.S. Department of Defense and certain industrial suppliers." The standards that do exist are workplace standards, Hinchey pointed out. They are not standards calculated on the possibility of people living and working with TCE who, potentially, could be exposed to TCE at some level twenty-four hours a day.

"What happened at the site is probably now lost in the mists of time," said Johnson at the hearing. New York State Assemblyman Patrick Manning, whose district includes East Fishkill—another site coping with VOC contamination leaked from an IBM facility and other industrial sources—found this response woefully inadequate.

"While the EPA debates what an acceptable level of TCE exposure is, the words 'draft guidelines' and 'gathering data' raise the hair on my arms, since it affects where I live," said Manning. "We've been told you can drink the water, you can shower in the water, but keep the window open so you don't breathe the vapors. How can you drink the water or shower without breathing the vapors? It's seems to me we're doing this on the fly. It makes people feel like they're being used as guinea pigs," Manning added.

In late 2004 Congressman Hinchey wrote a letter to the New York State Department of Health, requesting that the state act quickly to set strongly protective indoor air standards for TCE exposure: "Frankly, I am not content to wait indefinitely while this matter is studied. The people of Endicott, Ithaca, and other places where TCE has been detected at current actionable levels deserve quicker action and it is unconscionable that they are not getting it. Your agency is in a position to act."

|||| ||| |

Residents of Endicott are not the only ones affected by this contamination. The neighboring communities of Johnson City, Hillcrest, and Endwell, as well as forty thousand residents of Union all use water from the same groundwater aquifer. Their water is currently considered safe to drink, but that may change as more is learned about the pollution and also as TCE safety standards become more stringent. And it makes me wonder what will happen if Mountain View water supplies become scarce and residents there must tap into their contaminated aquifer—and about communities overseas where similar pollution may have occurred, particularly those with less stringent environmental oversight and less experience of citizen activism.

"It wasn't until we purchased a home in Endwell and looked at our water bill that we discovered we get our water from Endicott," Donna Lupardo tells me. She has lived in the area for thirty years and was elected in 2004 to represent Endicott in the New York State Assembly. "What people want to know is, am I safe in my home?"

"We use bottled water for drinking," Donna Turnbull tells me quietly, standing in her immaculate kitchen where a pot of coffee and plate of donuts have been set out to welcome visitors. "But I am concerned about showering, and I used to exercise in the basement," which is where TCE vapors could be entering the house, "but I don't do that anymore."

New York State Department of Health officials have stated repeatedly that the levels of volatile organic compounds detected in the Endicott water system are at levels below federal and state safety levels. But to many residents this provides little reassurance. They have begun to wonder if some of their neighbors' illnesses—like those of Bernadette Patrick's daughter who was diagnosed with Hodgkin's lymphoma at age seventeen, and residents of East Fishkill and Hopewell Junction who have become ill while living with TCE and related contaminants—may be related to the pollution. It's especially little comfort to hear the Department of Health say there is "generally no risk for people exposed to the pollution for short periods," but that "people who grew up in a house within the plume, or lived there for a long time, probably should check with the Department of Health to get more information about . . . long-term studies, regardless of whether they're ill."[48] These instructions are of little help to Patrick, who lived over the plume while she was pregnant. "What's worse, knowing or not knowing?" she asks rhetorically in her statement at the November 15 hearing. "The fear level is the same for everyone."

"Health issues are the number one concern," said one longtime Endicott resident I spoke to who asked not to be named. A perusal of the postings on the e-mail bulletin board established by the *Binghamton Press and Sun Bulletin*—the local newspaper that has reported conscientiously on the Endicott contamination and all its ramifications—reveals wrenching health concerns. The concerns include cancers, autoimmune diseases, neurological disorders, skin problems, and allergies among children and adults, as well as birth defects and pregnancy problems. These reports are, of course, purely anecdotal and the problems could stem from any number of causes, but the starting point for such studies is the gathering of anecdotal, self-reported information.

In 2003 the New York State Department of Health (DOH) began the first study to determine what, if any, of Endicott's residents' health problems might be connected to TCE exposure. The DOH released the first iteration of the study in the summer of 2005. The study showed that rates of low birth weights, infant heart defect, and of testicular cancer among residents living above the contamination plume are two to three times what would be considered normal. Further study is now being done, but the initial study results concluded that such rates of health impairment could not be considered entirely coincidential to chemical exposure.[49]

What makes Endicott residents particularly concerned about the health effects of TCE and other chemicals used in IBM's manufacturing is that longtime residents have had several potential routes of exposure. In 1987 IBM's Endicott plant was the United States' largest single source of ozone-damaging chemicals, releasing a reported 2.6 million pounds of chlorofluorocarbons and related chemicals. As recorded by the EPA's Toxics Release Inventory, among these chemicals were methyl chloroform and Freon. These are also among the chemicals listed in spills that ended up in storm sewers at IBM's Endicott plant. According to TRI figures, IBM released 1.6 million pounds of methylene chloride into the air in 1987. And while the company reduced its overall release of ozone-depleting chemicals in 1988, methlyene chloride emissions grew to 1.9 million pounds in that year. "Methylene chloride numbers are up because demand for printed circuits went up," Joseph E. Dahm, a spokesman for IBM told the *Press and Sun Bulletin* in 1989.[50]

A look at TRI reporting for IBM's Endicott plant from 1988 on shows that between 1988 and 1993, the plant's air emissions of toxic chemicals were substantial. In 1988 total air emissions were about 4.5 million pounds. In addition to methylene chloride these releases included 1.01 million pounds of Freon, over 1.1 million pounds of TCA, and nearly 300,000 pounds of other solvents, among them ethylene glycol, formaldehyde, tetrachloroethylene, xylene, and methyl ethyl ketone—all chemicals with serious health hazards.[51]

Ethylene glycol, Freon, TCA, methylene chloride, tetrachloroethlyene, and xylene, continued to be released annually through 1993. In 1994 IBM reported no releases of Freon, ethylene glycol, or TCA. In 2001 the plant reported about 21,500 pounds of air emissions, and in 2002—the year the property was sold to Huron Associates—air emissions were reported at zero. Curiously, not one TRI report between 1988 and 2002 (the most recent year available as of 2004) reported any releases of TCE to either air or water.

This data seems to back up Donna Lupardo's assertion that any health studies in Endicott should consider not only the quality of indoor air but also the quality of the outdoor air to which village residents and those who worked in Endicott were exposed. It also points to the many variables a study would have to consider. What might exposure levels be if an Endicott resident lived in a home above the contaminated groundwater plume, used village water, and worked at the IBM plant? How might air exposure levels vary depending on where you lived or worked in town? What time of year and in what kind of weather were the emissions released, and how might that have affected their impact on residents? And these are questions that would have to be asked even before questions of individual medical history and behavior patterns were considered.

While Endicott residents worry about the results of the air monitoring, they are also concerned that should they want to move, they will have difficulty selling a home located over the contamination plume. After months of negotiations with IBM—conversations in which New York State Attorney General Elliot Spitzer took part—and contemplated lawsuits, in September 2004 IBM came up with an offer. The company offered to pay Endicott residents whose property had been officially deemed affected by the groundwater contamination $10,000 or 8 percent of its value, whichever was higher. The offer had a deadline of November 30, 2004, and came with the condition that residents accepting the offer could not bring any future claims of property damage against IBM. They could, however, bring suit for personal injury or illness related to the contamination. (Personal injury is, in such cases, much more difficult to prove than property damage or loss of property value.) Many houses within the contamination plume are about seventy-five years old and have a market value of $40,000 to $70,000, reports the *Binghamton Press and Sun Bulletin*.[52]

One longtime Endicott resident who asked not to be named told me that she and her husband had decided to accept IBM's offer for a business property they own downtown that has a vapor intrusion ventilation system. "I feel like a turncoat doing this," she told me, "but my husband is seventy-seven and what we've accepted is less than fifty percent of the appraisal price, but we decided to cut our losses." But she added, "Making a decision like this about a business property is quite different than making a decision like this about a home. It's particularly hard for younger families who recently purchased homes at optimum prices who intended to be here for a long time. Elderly couples, if they're still here, they might take the offer."

She and several other residents said it has been difficult to get IBM to come back and take new readings on the TCE air levels, adding more questions about the health and safety information they are receiving. The frustration of uncertainty and underlying worry was clear in her voice. "It affects the psyche of those who live here and do business here. The whole area is stigmatized. This community has to worry about its tax base," she said. But Endicott mayor Joan Hickey Pulse remains positive about the town's prospects, as does the regional chamber of commerce. In the end, about 245 property owners accepted IBM's buyout offer, and the Village of Endicott accepted $50,000 in compensation for three vacant lots that are known to be contaminated.[53]

The fallout from the groundwater contamination is, in some respects, yet another body blow to an industrial town down on its luck. "I grew up in this town in the glory days," says Wanda Hudak. "And we didn't know anything but glory days until about 1985." IBM's steady employment kept the town relatively immune from the downturns other manufacturing towns suffered during the early 1980s.

"What's unique about this area," explains Donna Lupardo, "is that it was a two-company town. There was Endicott Johnson and there was IBM, which was born here. For the longest time, people couldn't cope with anything critical of these companies. A lot of people said, don't make waves."

Now Endicott is barely a one-company town and the waves have been made. "As far as TCE contamination is concerned," says Alan Turnbull, "Endicott is ground zero for the United States." No matter what happens next, it will take years to clean up the groundwater contamination that resulted from years of effluent coming out of the town's high-tech manufacturing industry. The one bright spot is that the situation in Endicott may push the local, state, and federal government to set more protective standards for exposure to TCE. It's also pushing local legislators toward enacting regulations that would—in the best of all possible worlds—prevent such a situation from occurring again.

"New York State and the state Department of Health need to revisit the issue of TCE standards," Assemblyman Tom DiNapoli told me. "It's important that we not compromise on issues of human health and that we have standards here in New York that are the most stringent of any in the nation. We've drafted legislation based on the precautionary principle, because I think it makes sense to take the conservative approach. We need the science, but we'll always be waiting for more science."

DiNapoli's comments go straight to heart of what has allowed so many toxic chemicals to be used—and used at great volume—without any real knowledge of their effect on human health and the environment. Solvents like TCE, ozone-depleting chemicals, and persistent bioaccumulative compounds have all been employed in high-tech manufacture (and other industries), all having been prematurely declared safe. And despite our knowledge of older chemicals' health effects, the system that made it possible for them to wind up in groundwater and the atmosphere is still in place, virtually ensuring that other persistent toxics will join them.

CHAPTER FIVE

Flame Retardants

A Tale of Toxics

"They're everywhere," Peter Ross a research scientist with the Canadian Institute for Fisheries and Ocean Science in Sidney, British Columbia, told me in May 2002. "They're all around us," said Dr. Arnold Schecter, professor of environmental sciences at the University of Texas Health Center, speaking on the phone from his office in Dallas in September 2004. "We're definitely eating them and probably inhaling a small amount," said Schecter.

You can't see them, smell them, or taste them, but if you live anywhere in the world where high-tech electronics like TVs, computers, cell phones, or CD players have been in use, it's almost guaranteed that they are in your home and office. And if you live in the United States and were to test the dust in your vacuum cleaner bag, the cheese in your lunch sandwich, the hot dog your brother ate at a baseball game last week, your breakfast eggs, or the farmed Atlantic salmon you grilled for the family dinner, chances are you would find them there too.

These unseen invaders are polybrominated diphenyl ethers (PBDEs), synthetic chemical compounds—they don't occur anywhere naturally— that are used as flame retardants in upholstery foam, carpet backing, textiles, and plastics, including those used in high-tech electronics. In ways that scientists are just beginning to understand, PBDEs are leaving the products in which they are used, making their way into the atmosphere, working their way through the food web, and showing up in our bodies. "You'll be chomping down on flame retardants pretty much no matter what you eat," said Ross.

PBDEs have been detected in the flesh and blood of marine mammals, fish, and shellfish in the Baltic Sea, the Gulf of Mexico, the Great Lakes, and in Greenland. PBDEs have been found in soil in the United Kingdom and Norway and in sewage sludge on the east coast of the United States. PBDEs have been found in polar bears living in the remote glaciers of northernmost Norway, in beluga whales and cod that swim the Arctic Ocean, and in orcas cruising Puget Sound. They have been found in peregrine falcon eggs from Sweden, in fish that live in a tributary of the Ebro River in Spain, and in sperm whales beached on the coasts of Denmark and Holland—an indication that PBDEs have reached the deep sea.[1] Levels of these chemicals found in herring gull eggs from the Great Lakes have been doubling every three years, while PBDE levels in San Francisco Bay harbor seals increased a hundredfold between 1988 and 2000.[2]

PBDEs have turned up in soil samples in southern China at levels ten to sixty times higher than found elsewhere in the world,[3] in food produced and purchased in Japan, in air samples taken in Taiwan, and in the food animals hunted and consumed by people living in the Russian Arctic. They have been found in the breast milk of nursing mothers across the United States and in the blood of all fourteen of the European government ministers who were tested for these compounds—people whose home countries ranged from Cyprus to the Czech Republic, Lithuania, and Italy. A 2004 study in the Pacific Northwest found nine women living near Puget Sound who had levels of PBDEs twenty to forty times higher than levels found in European and Japanese women.[4] Levels of PBDEs in Australians tested were five times higher than those found concurrently in Europe.[5] A family in Oakland, California, was found to have blood levels of PBDEs three to twenty-three times higher than the estimated U.S. average, which is ten to a hundred times higher than anywhere else in the world yet tested.[6]

"People do seem to care about what's in their bodies more than what's in the air or in the soil or even food," Clark Williams-Derry, research director for the Seattle-based nonprofit Northwest Environment Watch told the *Eugene Register Guard*.[7] And care they should, given that the National Institute of Environmental Health Services' journal has called the levels of PBDEs found in people "alarming" and that products we use daily expose us to a host of little-understood synthetic chemicals.[8]

Concerns raised by the rapidly escalating volume of PBDEs that's occurred in recent decades include their persistence in the environment and their potential for endocrine disruption and for neuro- and developmental toxicity, and thus their potential to affect children's health, particularly behavior and learning. PBDEs are "an environmental challenge," says toxicologist Thomas A. MacDonald of the California Environmental Protection Agency's Office of Environmental Health Hazard Assessment, who has been studying the health impacts of these compounds.[9] MacDonald's colleague, Kim Hooper, suggests that PBDEs may be the "PCBs of the future."[10]

I first heard about PBDEs at conferences I attended in the spring of 2002. The heated discussions—and sometimes shouting matches—between scientists, bromine industry representatives, European and U.S. environmental policy makers, fire-safety officials, and electronics manufacturers, left no doubt that the environmental impact of PBDEs was a controversial subject. While scientists were finding rapidly increasing levels of PBDEs in human and other animals' body tissue, the bromine products industry was asserting—as it continues to—that its products are safe and beneficial. In the interest of erring on the side of caution, Europeans were beginning to regulate and halt the use of some PBDEs, and high-tech electronics manufacturers were starting to follow their lead.

At the time, few people outside the world of science, toxics and fire-prevention policy, or electronics and chemical manufacturing had heard of PBDEs. Intrigued by these compounds that were turning up far from where they were manufactured or used, I wanted to find out how the health and environmental impacts of something so widely used could be so little understood. Several magazine editors I queried about writing articles on the subject responded, "Why would our readers care? It seems awfully obscure." Others asked if there was a smoking gun or a dead body anywhere. Several years later, as scientific investigation of PBDEs has burgeoned, newspaper articles seem to appear almost weekly.

The story of PBDEs seems emblematic of how the high-tech boom and our headlong embrace of "new and improved" products has encouraged a system that allows newly synthesized chemicals and other such materials to be used liberally, despite little or no knowledge of how they may affect human health and the environment. High-tech electronics alone aren't to blame for the spread of PBDEs and other brominated flame retardants, but without high tech it's highly unlikely that such large volumes of these materials would be in use and that so many products containing them would be spread throughout the world.

HOW PBDES WORK

PBDEs belong to a class of flame retardants known as brominated flame retardants (BFRs). They are halogenated compounds, meaning that they contain a halogen, in this case, bromine—one of a group of highly reactive nonmetallic elements that also includes chlorine, fluorine, iodine, and astatine. Chemical flame retardants of the BFR family work by interfering with one or more of the components needed for combustion—heat, oxygen, or fuel. When products containing BFRs reach very high temperatures, they slow combustion by releasing bromine atoms that starve the surrounding air of the oxygen needed to sustain a fire.

Brominated flame retardants do not stop a piece of equipment from catching fire altogether, but they sufficiently delay what fire professionals call "flashover" (when something bursts into total flame), thus allowing people to escape the impending blaze. To demonstrate the effectiveness of BFRs, the bromine industry has produced a video comparing the times it takes for a television with and without flame retardants to go up in flames. According to the Albemarle Corporation, one of the leading manufacturers of retardants, BFRs give "fifteen times more escape time" than non–flame-resistant plastics.[11] And, says the bromine industry, due to a lack of flame retardants, TV sets burn a hundred times more frequently in Europe than they do in the United States.[12] There's no question that fewer televisions and computers bursting into flames means fewer releases of polycyclic aromatic hydrocarbons, dioxins, and furans. But based on the evidence of harm caused by PBDEs, it's abundantly possible to argue that their use may not be the best way to achieve fire safety.

The bromine industry has written that Greenpeace wants to phase out all flame retardants because they are "a threat to life."[13] But a read through the environmental literature indicates that Greenpeace and its partner organizations would like to see BFRs phased out and replaced with alternatives that provide adequate fire safety. "To suggest that the environmental community is against fire safety is ridiculous," commented Robert Varney, New England regional EPA administrator at a 2002 conference on BFRs, although "it's equally wrong to suggest that advocates of fire safety are anti-environment."[14]

There are about seventy-five different kinds of brominated flame retardants, over half of which are used in electronics—in the plastic housings of computers, TVs, printers, fax machines, cell phones, cables, power sources, and the like, as well as in circuit boards.[15] Of all the flame retardants on the worldwide market—a business valued at over $2 billion annually and to which PBDEs contribute about $774 million[16]—BFRs are the most widely used. According to the Albemarle Corporation, "Hundreds of millions of pounds of brominated flame retardants are used every year because they are more compatible with modern manufacturing processes than any other compounds."[17]

PBDEs are but one type of brominated flame retardant, and several different kinds of PBDEs are used commercially. Of all the BFRs used in electronics, at least 50 percent are either decabromodiphenyl ether or tetrabromobisphenol A. However, many finished products containing PBDEs are in use or exist somewhere in the waste stream, where they continue to contribute to the flame-retardant chemicals being found in the environment, in wildlife, and in people.

The chemical properties of PBDEs make them particularly suited for use in hard plastics, including the widely used high-impact polystyrene (HIPS), acrylonitrile butadiene styrene (ABS), and polybutylene terephthalate (PBT)—all of which are common in electronics. PBDEs enable these kinds of plastics to meet high flame-resistance standards and to retain the properties that make them durable and easy to shape and color. From a manufacturing point of view, PBDEs are efficient and also happen to be the least expensive flame retardants currently on the market.[18] Consequently, PBDEs are "the most popular, most important part of the flame retardant business," Ronald Hites, director of Indiana University's Environmental Science Research Center, told me in August 2004.

▌▐║▌▐ ▌

Three kinds of PBDE compounds are used commercially in consumer products: pentabromodiphenyl ether (penta-BDE), octabromodiphenyl ether (octa-BDE), and decabromodiphenyl ether (deca-BDE), each distinguished by the number of bromine atoms each compound contains. The smaller PBDE compounds have been thought to be the least stable and have the greatest toxicity.

Penta-BDE is used mainly in upholstery foam and carpet liners, but it has also been used in circuit boards. Octa-BDE is used primarily in certain plastics and was also used in earlier generations of computer monitors, keyboards, and other electronic components. Deca-BDE, which Raymond Dawson of the Albemarle Corporation called "the workhorse" of the compounds, is the most widely used PBDE.[19] About 80 percent of all deca-BDE goes into the plastics commonly used in high-tech and other electrical appliances.[20]

While penta-, octa-, and deca-BDE are the commercially used versions of PBDEs, up to 209 versions of the PBDE molecule may exist. Each version, called a congener (related substance), appears to have a different potential for toxicity and is based on the number of bromine atoms the molecule contains and where the bromines are located on the molecule's two six-atom carbon rings. The specific commercial formulations of penta-, octa-, and deca-BDE vary depending on the manufacturer and on the year of production and may consist of more than one congener. When PBDEs turn up in the environment—in food, in animal tissue, in blood, breast milk, household dust, or soil—they are identified by what type of congener they are. Tracking PBDEs by congener provides the chemical identification card that helps scientists trace the compounds back to their origins in commercially used flame retardants.

The bromine business is a fairly concentrated one. Only a few companies manufacture bromine products and only a few countries extract most of the world's bromine supply. The United States and Israel together produce nearly 90 percent of it. Close to half is produced by a few U.S. companies from deep brine wells near oilfields—primarily in Arkansas, but also in Michigan—while Israel produces about 40 percent of its bromine from the Dead Sea.[21] About ten other countries also have active bromine deposits. Bromine can also be extracted from seawater and can be recovered by recycling sodium bromide, a chemical used in a wide variety of industrial and pharmaceutical applications and products. According to the U.S. Geological Survey, the world's supply of bromine is virtually unlimited, and says the Mineral Information Institute, "There is literally more bromine available cheaply than could ever be consumed at current rates, for many decades to come."[22]

Interestingly, brominated flame retardants were being developed and began to be released onto the market and used in large volumes just as the bromine industry's former major products were discovered to be ozone depleters. These included a bromine formulation used as an additive in leaded gasoline that released methyl bromide and a methyl bromide gas used as a fungicide (particularly on strawberries). The gasoline additive was phased out when leaded gas was taken off the U.S. market in the 1980s. (A number of other countries have also discontinued the use of leaded gasoline, but many still use it.)

The use of methyl bromide as a fungicide began to increase steadily in the early to mid-1980s,[23] but its use was curtailed by the 1987 Montreal Protocol, under which signing countries—including the United States— agreed to phase out use of ozone-depleting chemicals. The United States has been seeking an exemption for methyl bromide ever since. The EPA had scheduled the methyl bromide fungicide to be phased out of production and importation (with some exceptions) by 2005, but the George W. Bush administration has been seeking exemptions from that directive as well. In any case, flame-retardant use remains the driving force behind the world's current—and increasing—consumption of bromine and has more than compensated for the phaseout of other bromine products.[24]

As of 2005 there were three major manufacturers of BFRs and other bromine products: the Albemarle Corporation, based in Louisiana; the Great Lakes Chemical Corporation, headquartered in Indiana; and the Dead Sea Bromine Group, based in Beer-Sheva, Israel. Albemarle is a descendent of Ethyl Gasoline Company, which began producing an "anti-knock" product (tetraethyl lead) for gasoline in the 1920s and '30s. To solve a problem of corrosion caused by tetraethyl lead, a compound called ethylene dibromide was added to the mixture. Like Albemarle, Great Lakes Chemical also produced tetraethyl lead and ethylene dibromide. When leaded gas began to be phased out, ethylene dibromide was sold for use as a fungicide. When its health risks proved unacceptable, ethylene dibromide was replaced with a methyl bromide fungicide. Both Great Lakes Chemical and Albemarle produce methyl bromide for use as a fungicide as well as a whole suite of flame retardants including tetrabromobisphenol A, the BFR most widely used in circuit boards and of which methyl bromide is a manufacturing by-product.

Like any other major industry, the multibillion-dollar bromine industry works to promote its own products and interests. There is a striking contrast between the bromine industry's findings about the environmental impacts of PBDEs and those of the many scientists at universities all around the world not allied with the industry. Interpretations of risks and benefits, however, are infinitely more subjective.

Brominated flame retardants may be effective and efficient, but they also have a problematic history. The family of BFRs includes polybrominated biphenyls (PBBs), compounds that were taken off the market in 2000 due to toxicity concerns.[25] Studies determined them to have a tendency to cause skin disorders, an ability to affect the nervous and immune systems, to cause kidney, liver, and thyroid problems, and possibly to be carcinogenic to humans.[26] Another BFR known as "Tris BP" was used in sleepwear, but it was discontinued in the late 1970s after it was discovered to cause mutations and to be toxic to kidney function. Like PBDEs themselves—and nearly countless other synthetic chemicals—these earlier BFRs were used in consumer products without prior wide testing of their impacts on human health and the environment.

"One of the most elusive things about PBDEs," said Robert Hale of the Virginia Institute of Marine Science's Department of Environmental Sciences, "is good production statistics. The industry is very tight-lipped about how much is produced."[27] What is known is that consumption has grown dramatically since the 1980s, escalating on a timeline that coincides with the proliferation of high-tech electronics.[28] Linda Birnbaum of the EPA notes that there were "rapid increases in levels of PBDEs toward the end of the 1990s."[29] In 1992 world PBDE consumption was estimated to be about 40,000 metric tons.[30] In 2001 that number was approaching 70,000 metric tons.[31] In 2003 the Wall Street Journal put that number at 239,000 metric tons, a quarter of which the Journal estimated were used in the United States.[32] Altogether it's estimated that recent production levels of PBDEs are two to three times what they were when high-volume production began in the 1970s.

"The current generation of plastic and petroleum-based products is more flammable than older metal and wood products. And products that don't burn quickly save more lives than products that do," says toxicologist Thomas MacDonald, explaining the widespread use of PBDEs.[33] Add in the rest of the BFRs used in 1999—primarily tetrabromobisphenol A, which is used in circuit boards, and hexabromocyclododecane (HBCD), used in polystyrenes, some of which is made into packaging and housing for electrical equipment—and you get a total of nearly 160,000 metric tons of BFRs consumed. In 2000 the EPA estimated that worldwide demand for BFRs was 330,600 metric tons.[34]

Because the United States has the world's highest standards for flame resistance, North America consumes over half of all the PBDEs used in the world.[35] Asia, with all of its high-tech manufacture, is the next highest consuming region and also the region where, as of 2002, the rate of PBDE use was increasing the most rapidly.[36] Europe, where these compounds have been regulated and consumers, wary of the health impacts, have been opting for products without PBDEs, now comes in last among major regional consumers of PBDEs.

ARE PBDES THE "PCBS OF THE TWENTY-FIRST CENTURY"?[37]

Since the 1970s PBDEs have been used in products found in nearly every home, workplace, and government, educational, and healthcare facility in the developed world. Despite the fact that we're living with PBDEs—and likely ingesting and inhaling them—we know surprisingly little about how they affect human health. "No human health studies have yet been done, and I'm a little bit surprised," Dr. Schecter told me, speaking from his office at the University of Texas School of Public Health. Scientists have been gathering information about PBDE presence in the environment and wildlife since the mid-1990s, and testing for PBDEs in humans took place in Sweden in the late 1990s. But, said Schecter, whose research has documented significant levels of PBDEs in store-bought food in the United States, "The first analyses of PBDE presence in human blood and milk samples from the U.S. were done in 2000."

As more and more products containing PBDEs have flooded the world market, ever-increasing levels of PBDEs have been found *outside* of the products in which they are used.[38] Because of their toxicity we've stopped using many persistent organic pollutants—PCBs, dioxins, DDT[39]—but PBDE levels have "increased exponentially since the 1970s," writes Ronald A. Hites, professor at Indiana University's School of Public and Environmental Affairs.[40] Just such an "exponential increase in PBDEs was found in Swedish breast milk tested between 1979 and 2000," reports Myrto Petreas, an environmental scientist at the California EPA.[41] And when it comes to public attention, says her colleague Kim Hooper, "Breast milk speaks more loudly than sediment."[42]

||||| || |

One of the mysteries of PBDEs is exactly how they're escaping into the environment. They're not coming out of factory smokestacks or automobile tailpipes. They're not being dumped into rivers from factory drains or seeping into aquifers from leaky storage tanks. Only about a handful of companies actually produce PBDEs and other brominated flame retardants, and there isn't sufficient release from these sources or from other industrial facilities (like electronics recycling operations) to account for where PBDEs are turning up. "Logically," says Robert Hale of the Virginia Institute of Marine Science's Department of Environmental Sciences, "we shouldn't be seeing them where we are."

When used in plastics, a brominated flame retardant can either be mixed into the resin (or foam) in what's called an additive process, or in a reactive process it can form a chemical bond with the resin mixture. PBDEs are additive flame retardants, meaning they're stirred into the polymer or resin but don't physically bond with its chemical structure. This provides the first clue of what enables PBDEs to leave finished products: the flame retardant may diffuse from the surface of the material to which it was added.[43]

At a conference in 2002 I picked up a 2001 Bromine Science and Environment Forum (BSEF) publication that claimed that only one kind of BFR—penta-BDE—had been found in the environment far from production sites. BSEF explained this by saying there "was a historic use of penta-BDE in non-flame-retardant application such as hydraulic fluids in mines and drilling," which was discontinued in the late 1980s.[44] Material posted on the BSEF Web site in early 2005 said that research by the German Federal Environment Agency found that "no emissions could be detected from any of the three main BFRs," one of which is deca-BDE. "All studies confirmed that consumer exposure from BFRs is negligible," says BSEF.[45] Although numerous forms of PBDEs have been found in human breast milk, in 2001 BSEF asserted that only penta-BDE had been found in breast milk.

As California EPA toxicologist Thomas MacDonald explains, PBDEs are being released into the environment by the "billions of point sources from all these products."[46] Dr. Schecter told me he and his colleagues found it easy to wipe PBDEs off computer cases using a filter paper dipped in a solvent. And far too many PBDEs—including deca-BDE—have turned up in indoor air, food, human blood, and breast milk samples to be accounted for by hydraulic fluid applications that ended in the 1980s.

One of the questions scientists are trying to answer is how the quantities of PBDEs found in the food web relate to the volumes put into consumer products or otherwise released into the global environment.[47] Where PBDEs in the environment are coming from, Hale told me, is "a little more elusive than folks anticipated. We thought they were coming from fish—like PCBs—a classic path," he explained. "But the odd piece is how they're getting into people. We have limited samplings in people and they haven't really been tied to fish consumption. So house dust samples are now being looked at." Since this conversation in 2002 the number of human and household samples has grown, and the link between PBDEs and dust is growing as well.

"The big question with deca-, which is not very volatile, is whether it will debrominate down to things like penta-," the clearly more toxic compounds, said Hale. "Some studies show deca- breaking down in UV light in the lab, but we don't yet have a smoking gun that points to something in natural conditions." Yet experiments done in which "deca- was fed to carp fairly conclusively show, in my opinion, that deca- does debrominate," he said. Ronald Hites isn't so sure about this. But he adds, "It's not proven *not* to happen."

▌▐ ▌▌▐ ▌

While scientists are trying to determine precisely how PBDEs may be entering the environment, they already do know quite a bit about how PBDEs behave once there and what these materials are *not* doing. Because of their chemical makeup, PBDEs don't dissolve or become diluted when they come into contact with moisture. Because they're soluble in lipids, however, PBDEs can be taken up by the different kinds of fats and fatty tissue found in animals.

PBDEs are relatively large molecules that resist breakdown by enzymes and they're slow to biodegrade entirely, which is why they tend to accumulate in plants and animals.[48] Since the 1970s, levels of PBDEs in human blood, milk, and body tissue have increased by a factor of a hundred.[49] And younger people tend to have higher concentrations than do older people, an indication that they have likely been exposed to more PBDEs for their years than have their elders.[50]

"We're surprised that they move through the atmosphere as quickly as they do," said Jon Manchester, a researcher in the University of Wisconsin's water chemistry program. The prevalence of PBDEs, he said, "would suggest lots of sources."[51] Like other researchers, Manchester explains that atmospheric transport of PBDEs takes place when particles, presumably of dust containing PBDEs, move with rain and snow.[52] PBDE molecules "move with air masses, about five miles per hour, so it doesn't take too long" for these compounds to travel, says Hites.

▌▌▐▐▐ ▌

"Are BFRs the new PCBs?" asks the Bromine Science and Environment Forum, rhetorically. "No. There is no possible comparison between these substances."[53] Despite this assertion, numerous scientists have written in peer-reviewed journals that PBDEs "chemically and toxicologically resemble PCBs" in their tendency to persist in the environment, to linger in animal fat, and in their ability to travel long distances in the air and to be deposited in places far from where they were released.[54]

Nearly "every environmental monitoring program conducted during the past decade has shown sharply increasing levels of PBDEs in wildlife," wrote Mehran Alee and Richard J. Wenning in *Chemosphere*. "These observations are particularly troubling since PBDEs . . . similar to dioxins and PCBs are highly lipophilic compounds* and readily bioaccumulate through the food web."[55] Robert Hale agrees. "Structurally, PBDEs are like PCBs," he told me. "And I would be absolutely amazed if they didn't interact additively with PCBs."

The bromine industry has responded to such research by saying that

---

* Lipophilic means having an affinity for a tendency to accumulate in fat.

"findings of any man made chemical in human blood do not themselves equate automatically with a health risk." In their view there isn't enough information available to determine whether levels of PBDEs detected in the global environment are in fact rising.[56] But according to Thomas MacDonald, "over the past two decades levels have been doubling every two to five years, which suggests we have an emerging problem."[57]

"If you look over the evolution of what's happening with PBDEs," Hale told me, "you see industry beginning to retreat a bit." But he added, "really what you're seeing is damage control."

It now seems impossible to deny that the stuff is out there and is making its way into at least some—if not many—of us.[58] Since the late 1990s, levels of PBDEs have begun to decline in Europe, where their use is being curtailed, but they have jumped up in the United States.

In one alarming finding published in early 2005, a twenty-month-old boy in Oakland, California, was found to have levels of PBDEs in his blood nearly three times higher than those at which scientists begin to see behavioral changes in lab rats.[59] And research published in 2004 made it quite clear that PBDEs are an unbidden part of the average American diet.

Tests of groceries bought in American supermarkets revealed that nearly all food of animal origin was contaminated with PBDEs—shrimp, ground turkey, chicken, butter, ice cream, eggs. Even soy formula was contaminated. Pork sausages, a salmon fillet, hot dogs, and cheese had the highest levels, and of all the food tested, only nonfat milk was PBDE-free. Compared to similar studies done in Spain and Japan, the U.S. food levels of PBDEs were significantly higher. The PBDE congeners found included those that make up the commercially formulated penta-BDE flame-retardant products, but also the congener that makes up the deca-BDE product, as well as others that may result from breakdown of penta-, deca-, and other PBDE products.[60]

Anyone who's been eating salmon over the past few years—just about everywhere in the world—has almost certainly consumed PBDEs, according to the results of a study published in 2004 by Hites and his colleagues. Following up on research that showed levels of persistent bioaccumulative contaminants—including pesticides, dioxins, and PCBs—to be higher in farmed salmon than in wild salmon, Hites and his colleagues decided to investigate whether the same might be true for PBDEs. It was. But wild salmon samples also contained PBDEs, some at remarkably high levels. This is of special concern because salmon, both wild and farmed, have become an increasingly popular and important food in recent decades (with ever greater amounts of farmed salmon being consumed as wild stocks dwindle).[61] "It has been suggested that PBDE concentrations now observed in humans may leave little or no margin of safety; thus, prudent public health practice argues for the selective consumption of food," write Hites and colleagues.

When I asked Hites exactly what these findings might mean, he told me there has "not been enough toxicology done on brominated ethers so we don't yet know at what level to be concerned, but the chemical structures are similar enough to PCB structures, and concentrations are going up which causes one to be a little bit concerned . . . Would I stop eating salmon personally? Probably not," he said. "But I'm an old guy. If I were a pregnant woman in the first trimester of pregnancy, particularly with other studies we've done factored in, I'd think twice about it. My coauthors don't want to eat salmon any more."

"What can we do right now?" I asked Dr. Schecter, whose study found PBDEs in a whole array of food. "Eat less animal fat and choose low fat or skim milk products," he told me, also suggesting that the best way to cook meat and fish is to "broil and drip it."

PBDES AND HUMAN HEALTH

"The toxicological endpoints of PBDEs are likely to be thyroid hormone disruption, neurodevelopmental deficits and cancer. Unfortunately, the available toxicological evidence for these endpoints is surprisingly limited, given the widespread use, bioaccumulative potential and structural similarity to thyroid hormones and polychlorinated biphenyls (PCBs)," wrote Thomas MacDonald in an article published in *Chemosphere* in 2002.[62] Research by MacDonald and numerous other scientists has shown that PBDEs, like PCBs, can act as endocrine disruptors and interfere with thyroid hormone function, particularly at the fetal stage. Thyroid hormones regulate metabolism and are essential to children's growth and nervous-system development, so an upset of this system may have many ramifications. PBDEs have also been shown to cause neurological and reproductive development problems in laboratory animals and to have an effect on how certain brain chemicals determine motor behavior.[63] Some of these neurological problems are similar to those linked to learning disorders, attention deficit disorder, and hyperactivity in children. Lab studies of mice and rats indicated that very high doses of PBDEs in animals may lead to cancer,[64] although to date, deca-BDE is the only PBDE that has been tested this way.[65]

Despite the growing evidence,[66] the bromine industry continues to say that the health and environmental risks of PBDEs are minimal and that that they in no way resemble PCBs.[67] "To date, no human health or environmental effects have been associated with BFRs," said the Bromine Science and Environmental Forum Web site in 2005. "There seems to be a misconception that BFRs, as a group, accumulate in animal and human tissues."[68] The BSEF statements also gloss over the potential impacts of the billions of units of electronic equipment out in the world and in the waste stream that are releasing PBDEs into the atmosphere. Yet, say Linda Birnbaum, Ronald Hites, Dr. Schecter, and other scientists studying PBDEs, much more research—including widespread epidemiological studies—needs to be done, particularly because PBDEs were launched commercially without any monitoring of how their use in consumer products might impact human health.

"THE PUBLIC, QUITE REASONABLY, GETS UPSET"

Now, some thirty years after their introduction, after many millions of tons have been used in consumer products, concern about the effect of PBDEs on human health and the environment has led to their regulation in Europe. Beginning in 2006 penta- and octa-BDE—whose potential for adverse health impacts are the best documented—will be barred under the European Union's Restriction on Hazardous Substances (RoHS) directive. Use of the third compound, deca-BDE—the PBDE commonly used in the plastics that house televisions, computers, and other electronic appliances—is not yet regulated anywhere. However, the EU regulations are written so that if and when health concerns arise, additional substances can be added to the list of those prohibited under RoHS. As of spring 2005, neither the U.S. nor Canadian federal governments had any nationwide restrictions on any PBDEs, nor did Japan or Australia.

Yet confronted with growing public concern, a number of electronics manufacturers (and other companies that make products containing BFRs) are discontinuing the use of some or all PBDEs in many of their products. A far from complete list includes Intel, Sony, Hewlett-Packard, Dell, Philips, IBM, Motorola, Panasonic, Samsung, Toshiba, and Apple. To avoid and reduce use of PBDEs, manufacturers are limiting the plastic parts where PBDEs can be used. Some are redesigning equipment so fewer flame retardants are needed, and some are substituting other flame retardants. For example, in December 2004 Hewlett-Packard restricted further use of a number of PBDEs in its products, a list that includes penta-, octa-, and deca-BDE. The company's policy restricts use of its listed PBDEs, PBBs (already out of use due to their toxicity), and polybrominated biphenyl ethers or polybrominated biphenyl oxides in "plastic parts, components, materials and products" in "concentrations greater than or equal to 0.1% (1000 PPM) by weight."[69]

Dell has similarly discontinued using brominated fire retardants in the plastic housings of many of its products. Instead the company is using plastics "that can be flame-rated with phosphorus-based flame retardants and by using design strategies that reduce the need to use flame-rated plastics at all." But Dell also says that some plastics "cannot be flame-rated with anything other than bromine because reliable alternative technology does not currently exist." Circuit boards—which are nearly all made fire resistant with another BFR, tetrabromobisphenol A—would be one such example. Yet, says Dell, "we try to avoid these types of plastics when we need flame retardancy."[70]

Apple's policy on flame retardants says that plastic enclosure parts that weigh over twenty-five grams will not contain any BFRs or antimony trioxide (another compound commonly used to make plastics fire resistant). While it "uses tetrabromobisphenol A (TBBA) as a flame retardant for printed circuit boards, which is standard across the industry," the company says it is "actively researching equally effective alternatives with better environmental features than TBBA."[71]

While high-tech electronics manufacturers and other companies that use flame retardants were beginning to shift away from some PBDEs, and with the EU ban on penta- and octa-BDE on the horizon, the Great Lakes Chemical Corporation announced that it would discontinue its production of penta-BDE by the end of 2004. Because Great Lakes Chemical was the only U.S. manufacturer of penta- and octa-BDE (the Albemarle Corporation didn't make penta- and stopped producing octa-BDE in the 1990s), as of 2005 neither product is made in the United States There are, however, plenty of products containing penta- and octa- currently in use, and will be for some time to come. Things have changed quickly in this arena, considering that in 2001 the Bromine Science and Environment Forum assured the public that "apart from one long-existing EU Directive which excludes the use of PBBs in clothing, there is currently no legislation restricting the use of individual BFRs."[72]

Despite "the hoopla, the phase out of the two products isn't as earth shattering as would be an exit from deca-BDE," said a plastics industry consultant commenting on Great Lakes Chemical's move. "They just agreed not to produce materials that are in decline or not produced at all," he told *Chemical and Engineering News*, noting that deca-BDE makes up about 90 percent of all PBDEs used in North America.[73]

Meanwhile, a number of U.S. states have begun to restrict sales of products containing PBDEs. Prompted by rising levels of PBDEs in the San Francisco Bay area, in 2003 California State Assembly majority leader Wilma Chan introduced a bill to phase out in-state sales of products containing penta- and octa-BDEs.[74] "When Washington is unable or unwilling to act, we must protect the health of our citizens," said Governor Gray Davis when he signed the bill, which allows continued use of deca-, but bans the other PBDEs beginning in 2008. "For the governor to say the EPA is taking no action, that's not true. We're moving toward getting the information we need from the industry while working with the industry to find alternatives," responded Mark Merchant of the EPA, speaking to the Associated Press.[75]

Since then, Maine and Washington state have enacted legislation regulating PBDE use that generally mirrors the European Union's. Maine banned penta- and octa-BDEs and will ban deca-BDE if an adequate substitute can be found by 2008, and Washington is considering such a ban that would proceed along a similar timeline. A number of other state legislatures have introduced comparable bills that would restrict PBDEs from products sold within their borders. "This is a common sense response to a serious public health threat," said state representative Hannah Pingree, sponsor of Maine's legislation.[76]

This kind of local action, Ronald Hites told me, is "largely as the result of the scientific literature getting to the public and the public saying, 'This doesn't make any sense,' a move often led by the Europeans, particularly the Swedish who use the precautionary principle, saying 'This looks a lot like PCBs, let's ban it,' as opposed to the U.S. who says, 'Let's see if it kills anyone before we ban it.'" He continued, "It's a tough issue to deal with in a large population with hugely confounding variables." But, he said, when faced with what we're learning about these compounds, "the public quite reasonably gets upset."

TETRABROMOBISPHENOL A: ANOTHER FLAME RETARDANT ON THE MOVE

While attention has been focused on PBDEs, investigation of another widely used brominated flame retardant is prompting questions about its environmental impacts. Tetrabromobisphenol A (TBBPA, sometimes abbreviated TBBA) accounts for over half the flame retardants used worldwide as of 2005. Current annual usage is estimated to be between 120,000 and 150,000 metric tons. Approximately 96 percent of all circuit boards currently manufactured contain TBBPA. Unlike PBDEs, this compound is a reactive flame retardant, meaning that it binds chemically with the plastics to which it's added. This makes it less likely to detach itself physically from finished products, yet TBBPA has been detected in the environment and in people and raises its own set of environmental and health concerns.

"TBBPA is not chronically toxic or mutagenic," says Linda Birnbaum. It degrades in soil and sunlight and, while it accumulates readily in fat tissue and is easily absorbed in animals, it also seems to be metabolized and eliminated quickly, Birnbaum explains, so that little is retained in tissues. Yet results from research done with laboratory animals is worrisome, she continues, because it has been discovered that TBBPA can be toxic to the immune system, can interfere with thyroid hormone function, and can be neurotoxic within cells.[77] It has also been shown to be very toxic to aquatic organisms and to accumulate in fish. While TBBPA is considered by the National Institutes of Health to be both persistent and toxic, it is not currently considered hazardous *enough* for the EPA to declare it a persistent bioaccumulative toxin

TBBPA has been detected in freshwater, in sewage sludge, and in sediments in Sweden near a plastics-industry facility, as well as in sediments and fish in Japan, and in landfill sludge.[78] It has also been found in fish and shellfish. Computer technicians, laboratory personnel, circuit board producers, smelter workers, and electronics dismantlers tested in Norway were found to have TBBPA in their blood.[79]

Outside the factory, tetrabromobisphenol A seems to be entering the environment primarily through industrial wastewater. Some has been detected in industrial air emissions, and a small amount seems to be released from products in which it's used.[80] Once in the atmosphere, TBBPA is likely to degrade and under certain conditions its breakdown products will include bisphenol A—a compound that has been used in many plastic products for decades (including polycarbonate water bottles)—about which there are longstanding health concerns. TBBPA itself is thought to be an endocrine disrupting compound that affects thyroid hormone function, while recent research indicates that bisphenol A can affect reproductive hormone and prostate gland development in mice, which has led scientists to suspect similar consequences in humans.[81]

The studies cited by high-tech manufacturers to tout the safety of TBBPA were done by the World Health Organization in 1995. These studies found that the compound had little potential to bioaccumulate and that its human health risks were insignificant. But more recent research has raised sufficient questions about TBBPA's safety for the Convention for the Protection of the

Marine Environment of the North-East Atlantic (the OSPAR Convention, to which EU countries are party) to place TBBPA on its list of chemicals "for priority action"—a decision that has launched an ongoing series of studies. The initial results of OSPAR's risk assessment, realeased in 2004, showed the need for continued "concern over some uses of tetrabromobisphenol-A as a flame retardant," and called for additional monitoring of TBBPA's environmental and health risks as well as a search for safer substitutes.[82]

But finding safe alternatives to chemicals like tetrabromobisphenol A—which itself was substituted for octa-BDE in circuit boards when concerns over its environmental and health impacts arose—is complicated by the way the Unites States regulates the use of chemical products. The U.S. system relies on manufacturers' assessments of their products' safety and focuses on the short-term impacts of high doses when assessing risks. "Penta- is now banned in the EU, in California and will be in Maine—and Great Lakes has gone to something called Firemaster 550, but the structure of that is proprietary," Hites said when I asked him about BFR manufacturers' response to the most recent assessments of their products' environmental safety. "I'd do the same thing," said Hites. "Behave like good guys, take a product off the market, replace it with another and get five to ten years out of it, and in practice that's what's happened."

Industry analysts have a similar view. "As new brominated flame retardant products are a point of differentiation between companies, manufacturers have had to invest in continuous research and development to sustain business growth. Many manufacturers of brominated flame retardant chemicals have already diversified into other flame retardant materials, in case more stringent and/or widespread legislation limiting the use of flame retardants comes into force," writes the Roskill Consulting Group.[83] But this evolution of chemical products often makes it even harder for the public to understand what is safe.

PRECAUTION, OR HOW DO WE LET THIS STUFF HAPPEN?

We buckle seatbelts, strap our children into car seats, wear bike helmets, and at airports let uniformed strangers look inside our shoes. Many of us make grocery purchases based on what a product is certified *not* to contain. Seatbelts, child safety seats, and helmets can protect us in case of an accident. Screening airplane passengers presumably protects us from those who might cause deliberate harm. Taking precautions to prevent physical accidents and crime is one thing, but how do we guard against hazards posed by things we cannot see or hear? It's one thing to accept the known risk that your children may fall down in the playground. It's another to accept the unbidden risk that the wooden equipment or plastic in their toys contains chemicals that may adversely affect their health.

To avoid things we are allergic to, and to make choices about what we consume and use on our bodies and in our homes, we scan lists of ingredients. But what about all the manufactured items that don't list their materials—household cleaners, office products, and electrical appliances we touch constantly, like phones and computers, and furniture, upholstery, and the clothing that hugs our bodies?

When much of what we buy is infused with chemicals that make things easier to wash, faster to dry, more colorful, harder to break, easier to bend, and resistant to germs, stains, and flames, these questions become pressing. According to the EPA about seventy-five thousand industrial chemicals are now produced in or imported into the United States. To enable the EPA to track, test, and regulate these chemicals, in 1976 Congress enacted the Toxic Substances Control Act. While new chemicals must be tested before being produced commercially, all chemicals on the market before December 1979—that is, most chemicals now in use—are considered safe unless they're demonstrated to present an unreasonable risk to human health or the environment. Initial testing of new chemicals is the responsibility of the producer, not of the government or an independent third party, which raises the obvious question of whose interests will be put first, those of the chemical manufacturer or those of the public?

▌▌▐▌▌▐ ▌

In September 2003 a study by the Silent Spring Institute in Massachusetts found fifty-two different chemical compounds in household air and sixty-six different chemical compounds in household dust samples taken from homes on Cape Cod.[84] All of the chemicals found are known to be endocrine disruptors—chemicals that interfere with normal hormonal activity, including reproductive functions and sexual development. Some have also been shown to disrupt neurological function in animals.

In addition to polybrominated diphenyl ethers, these chemicals included phthalates that are used in plastics—like those in children's toys—and nonylphenols used in disinfectants, detergents, plastics, adhesives, and pesticides. Phthalates have also been used in plastics that wrap and contain food; their residue has been detected in food samples, as have nonylphenols. Some of the chemicals found in the Silent Spring Institute study had been off the market for more than twenty years.

In November 2003 the World Wildlife Fund published a study done in England that sampled 155 volunteers' blood for evidence of persistent bioaccumulative chemicals.[85] Every sample tested contained more than one such compound, including PCBs, the flame retardants used in the plastics found in electronics and in upholstery, pesticides, and remnants of DDT. None of the volunteers worked with these chemicals in a manufacturing or industrial facility. Moreover, PCBs and DDT have been banned for decades, so this study presents new evidence of their persistence in the environment.

Some forty years after the publication of *Silent Spring* we know that chemical products developed to do a world of good can also cause great harm. How the costs and benefits of such products are weighed, and how we decide to act after assessing their risks, are central to the debate over what's called the precautionary principle.

What actually is the precautionary principle? "There are a zillion definitions out there," George Gray of the Harvard Center for Risk Analysis told me in 2003. While wording may vary, the prevalent understanding is that "precaution is about anticipating and preventing environmental health damage before it occurs. It is about using all the available evidence on hazards and alternatives to make the best possible decisions that prevent harm to human health or the environment," writes Joel Tickner of the University of Massachusetts Lowell's Department of Work Environment.[86] Or as ecologist Sandra Steingraber says in her book *Living Downstream*, the precautionary principle "dictates that indication of harm, rather than proof of harm, should be the trigger for action—especially if delay may cause irreparable damage."[87]

The European Union's legislative body, the European Commission, uses the precautionary principle in its environmental policy making and calls the principle a decision-making tool to be used "when we are faced with potentially harmful effects on health or the environment, but there is scientific uncertainty concerning the nature or extent of the risk."[88] The precautionary principle was at work in creating the European Union's RoHS directive, but in the United States the federal government currently finds the concept somewhat anathema. For example, in a January 2002 speech John D. Graham, administrator of the U.S. Office of Management and Budget's Office of Information and Regulatory Affairs, said that "the U.S. government supports precautionary approaches to risk management but we do not recognize any universal precautionary principle. We consider it to be a mythical concept, perhaps like a unicorn."[89]

So why did the City of San Francisco pass an ordinance in July 2003 making the precautionary principle integral to the city's environmental policy? As Mayor Willie Brown's office put it, "The Precautionary Principle maintains in essence, 'An ounce of prevention is worth a pound of cure' . . . If a practice poses a threat to human health or the possibility of serious environmental damage, the Precautionary Principle approach will use the best available science to identify cost-effective alternatives."[90] San Francisco's ordinance is no rainbow chase. Its immediate goal is to eliminate and find safe alternatives to toxic chemicals used in the city's cleaning and maintenance operations. With similar proactive goals, the City of Seattle passed a comparable resolution in July 2002 to assess, reduce, and find alternatives to persistent bioaccumulative toxins used in city offices and operations (e.g., in cleaning solutions, pesticides, plastic products, and upholstery foam).

A bill based on the same approach was introduced in the Massachusetts state legislature's 2005 session. Called the "Act for a Healthy

Massachusetts," the bill, which has been introduced twice before, would phase out and find safe alternatives to ten chemicals used widely despite their toxicity and growing scientific evidence of their adverse impacts on human health and the environment. In the summer of 2005, the legislature funded the first step toward this goal: a study of five of these toxics.[91]

The EPA supports toxics reduction, so why the federal government's objection to policies that invoke the precautionary principle? Because the precautionary approach to chemical safety allows for action in the face of uncertainty, it's "a bit of a threat to the way things are right now," suggests Daryl Ditz of the World Wildlife Fund.[92] The U.S. approach to chemical safety has been to assess risk, determine what level of risk is acceptable, and to wait for definitive proof—or as close as science can get—that a substance will cause harm before declaring it unsafe. Given the way chemicals act on the human body it's often difficult to chart a direct cause and effect, especially since sensitivity and reactions to toxins vary from person to person. When regulatory action isn't taken until discrete damage is proven, it's difficult to take chemicals out of use even with evidence of harm to ecological systems—including the human body.

This antiprecautionary stance, which allows the continued use of hazardous chemicals, is clearly of benefit to those who manufacture and profit by their use. This status quo will change if the European Union passes its REACH legislation (Registration, Evaluation and Authorization of Chemicals), which will require registration of all chemicals used in volumes over one metric ton, along with publicly available information about their toxicity, health, and environmental impacts. Chemicals lacking this data will not be allowed on the European market. REACH would apply to imported and EU-produced chemicals. The U.S. government has been lobbying hard against REACH, arguing that this legislation could, as former secretary of state Colin Powell wrote in a cable sent to U.S. embassies in Europe, "negatively impact innovation and EU development of new, more effective, and safer chemicals and downstream products."[93]

"The precautionary principle is a red flag thing," Joel Tickner told me.[94] "Say it and you have industry at your door calling you antiscience and anti-innovation." The George W. Bush administration has taken this stance, calling the precautionary principle a potential barrier to the marketing of new products and technologies. Tickner counters that the search for safer alternatives to currently used toxics will require serious scientific assessment and will spur innovation, research, and development.

Given the ongoing use of large numbers of chemicals with demonstrable toxicity, public pressure is increasing worldwide to protect health by stemming the flow of such chemicals, rather than by solely determining safe levels of exposure. Absent U.S. federal regulation in this area, states and local governments are beginning to act. There is "broad concern that current policies aren't protecting health sufficiently," Cynthia Luppi, organizing director of Clean Water Action, told me.[95] "Almost every family in the state [Massachusetts] is somehow personally concerned about the exposure and buildup of the chemical soup in our bodies," she said. "It's time to look at safer alternatives to the current system."

A world without risk of chemical exposure may be utopian, and decisions about risk are personal, but there is growing consensus—and scientific evidence—that continuing to allow hordes of chemicals to infiltrate our bodies and the world's wildlife is not in the public's best interest.[96] And, as Joel Ticker points out, "A lack of information should not be interpreted as safety."

"THESE THINGS AREN'T GOING TO LEAVE ANYTIME SOON"

In December 2004—despite the apparent federal antipathy toward action on precaution—the EPA announced a proposed rule under the Toxic Substances Control Act that would require manufacturers or importers of several types of PBDEs (a list that excludes deca-BDE) "to notify EPA at least 90 days before commencing the manufacture or import of any one or more of these chemical substances on or after January 1, 2005, for any use. EPA believes that this action is necessary because these chemical substances may be hazardous to human health and the environment. The required notice would provide EPA with the opportunity to evaluate an intended new use and associated activities and, if necessary, to prohibit or limit that activity before it occurs."[97] Since no one is making penta- or octa- in the United States, and most manufacturers who use PBDEs and BFRs in their products have moved away from these compounds, this rule is unlikely to have much impact on either the flame-retardant or electronics industries.

In May 2005, at its meeting in Uruguay, the United Nations Environment Programme announced that it is considering adding penta- and hexa-bromodiphenyl ether to its list of chemical compounds that would be phased out globally under the provisions of the Stockholm Treaty on Persistent Organic Pollutants. Over 150 countries have signed and close to a hundred have ratified the treaty. And although the United States—the world's largest consumer of PBDEs and home to two of the world's major manufacturers of brominated flame retardants—has signed on, it has not yet ratified the treaty.

While these slow steps are being taken, concern has been growing over the health impacts of simultaneous exposure to multiple toxic chemicals. "Looking at mixtures is fairly new," Dr. Schecter told me. "But we do know that dioxins, furans, PCBs, mercury, lead and likely PBDEs, all are additive," meaning that concurrent exposure to more than one of these chemicals can magnify or otherwise increase their impacts. "Current government safety standards are for single chemicals. They assume no other chemicals are present. Monitoring for multiple chemicals is something we've got to do," he said.

As we wait for further health studies, more concerted testing and monitoring, and swifter action in response to hazards posed by chemicals used in high-tech electronics and so many other products in daily use, these toxins continue to swirl around us and inside us. And in a strange twist of timelines, it seems that many of the chemicals that enable faster-paced and more flexible lifestyles often outlast the products that contain them. As Dr. Schecter put it, "These things aren't going to leave anytime soon."

CHAPTER SIX

When High-Tech Electronics Become Trash

The scenario is familiar. The day comes when the computer that was going to be your personal bridge to the twenty-first century has become a dinosaur. The salesperson who touted that machine's efficiency now explains in tones of pity and derision just how far from the cutting edge of new technology you are. Or you see an advertisement from your "wireless service provider" announcing a new bargain calling plan—available only with the purchase of a new phone, which, thanks to rebates and discounts, will cost you nothing. So, although your current cell phone works just fine despite a little crack in the case, you get a new one.

In early 2001, when it became clear that my old laptop couldn't handle most Web sites and could not be upgraded, the only solution was a new computer. I tried to find someone who wanted my still functional Macintosh, but no one I knew was interested in a computer that couldn't surf the Web without crashing, so into the closet it went, along with my old cell phone, a defunct cordless phone, and an old zip drive.

I knew my trashcan was not an environmentally responsible place for an old computer, so I called some local electronics retailers for advice. One suggested I sell it on eBay.* Another said donate it to a school or give it to

---

* In 2001, this seemed like a joke. But in January 2005 a program to put computers into reuse and to facilitate recycling of high-tech equipment was launched by eBay, and the Computer TakeBack Campaign. However, finding someone who can use a computer that can't run current software, handle current Web sites, or be upgraded remains a decided challenge.

charity. None had any provision for recycling. One store said just put it in the trash. Apple itself didn't then have a program to collect individual machines, so I called the company that picks up the garbage and recycling bins in my neighborhood. They didn't collect electronics for recycling, but suggested I call Metro, the regional government in the Portland, Oregon, area that regulates local waste disposal. Metro gave me the names of several private recycling companies located in the suburbs that, for a fee, would accept certain electronics products. Eventually I gave it to a friend who knew someone who could use it.

There are now so many pieces of abandoned high-tech electronics stashed in private basements and closets that some recycling professionals refer to them as "closet-fill." Add to this the used high-tech equipment coming out of large businesses, institutions, and government offices and you get what Michael Paparian of the California Integrated Waste Management Board called an "e-waste stream" that "is growing at an alarming rate both in quantity and complexity."[1]

THE "SKYROCKETING RATE OF OBSOLESCENCE"

How much high-tech trash is out there? What happens to this equipment when those who have purchased it no longer find it useful? This may seem like elementary information, but only in the last decade or so have any significant attempts been made to quantify the extent and distribution of e-waste. Indeed, the whole idea of quantifying trash and considering the end of a product's life at, or before, its beginning is a relatively new concept. In the United States, where we still cling to the myth of the endless frontier and equate progress and prosperity with the ability to jettison something old, the notions of reuse and recycle have been particularly slow to take hold. American businesses tend to regard the idea of producer responsibility—which holds the manufacturer responsible for a product at the end of its useful life—as a threat to profit margins. In Europe and Japan, however, space—both public and private—is more limited than in the United States, so it's harder to stash or toss the trash without considering the implications. Consequently, citizens of those countries have become more comfortable with regulations that accommodate these conditions.

The first step in figuring out how much e-waste is out there is to find out how much high-tech stuff there is to begin with. The numbers are mind-boggling, but they're fundamental to assessing the problems posed by e-waste—and they're not a matter of abstractions. When reading about them, try to picture what these numbers represent in terms of the space this equipment occupies, its weight, all the peripheral parts, interior components, and the many materials each device contains.

Also, consider that computers, televisions, cell phones, and their ilk are unlike many of the items we regularly recycle (newspapers, cans, and bottles, for example). Given the current design and materials content of high-tech equipment, electronics can't be safely broken up or disposed of except under professionally controlled conditions.

Add to all this the complication that plastics, used most extensively in high-tech products, have the environmentally perplexing quality of being perceived of as disposable while they are, for the most part, stubbornly durable. This durability can become an environmental liability. One plastics scientist has quipped that the plastic bag will outlast not only your sandwich, but will likely outlast you as well. Plastics—of which there are many different kinds (often in one piece of equipment), many containing toxic additives—have proven to be the most difficult materials in high-tech electronics to recycle. This variety and toxicity have contributed to the challenge of finding a profitable market for post-consumer plastics. And because of their chemical composition, disposing of plastics from used electronics in landfills or municipal incinerators is not environmentally sensible (although, alas, there is sometimes no other choice).

▌▐║▐▌

In 2005 there were approximately one billion personal computers and well over a billion cell phones in use worldwide.[2] The greatest concentration of high-tech electronics resides, not surprisingly, in the world's richest countries, but it's only a matter of time until the world's most remote and poorest nations gain on countries already awash in high-tech electronics.

As of this writing, over two hundred million of the world's computers are in the United States.[3] With over five hundred PCs per thousand people, the United States has the highest per capita concentration of these machines of any large country. Altogether, there are now an estimated two billion consumer electronics products in American homes and small businesses— a number that does *not* include the computers in corporate and government offices or in educational, health care, and other institutions.[4]

The next most PC-populous regions are Europe, Canada, Hong Kong, Japan, South Korea, and Australia, which all average about two hundred to five hundred PCs per thousand people. In northern Europe the concentration of computers approaches that of the United States, and matches it in Scandinavia. Moving south and east, the number of PCs decreases to about fifty to two hundred per thousand people in Spain, Portugal, and eastern Europe.[5]

India, the world's most populous democracy, is home to a burgeoning high-tech industry and is the site of many American companies' off-shore call centers—which means a lot of computers and telecommunications equipment. Nationwide, India now has no more than about ten PCs per each thousand of its over one billion people. But India's high-tech sector is estimated to be growing at a rate of about 40 percent a year,[6] and the country— already burdened with e-waste exported from richer countries—currently produces about 1,050 metric tons of domestic electrical scrap as well. Meanwhile in China, which has experienced a stunning rate of industrial and commercial growth in the past decade and which has the dubious distinction of being the world's largest recipient of other countries' e-waste, there are at least ten to fifty PCs for each of that country's 1.3 billion people.

Americans may own more pieces of high-tech electronic equipment than people in any other nation, and it may seem like everyone in the United States is chattering on a cell phone, but mobile phones (as cell phones are known outside the United States) are more ubiquitous elsewhere. Europe, Japan, Israel, Australia, South Korea, and New Zealand all exceed the United States (and Canada) in per capita ownership of these devices. In these countries, there are now some five hundred to a thousand *or more* mobile phones per thousand people, about twice the density of cell phones in the United States.[7] This gives the United States about the same concentration of cell phone ownership as Mexico, Argentina, South Africa, and Saudi Arabia—just to give a sense of the ecumenical nature of potential e-waste. While China's overall ownership of phones is but about fifty to two hundred per thousand people, in Beijing and other commercial centers, cell phones are rampant. On a visit to Beijing in April 2004 I saw far more mobile phones in action than I ever have in a comparable stretch of time in New York.

As for televisions, the United Nations puts U.S. per capita ownership as the world's highest. According to the 2002 World Radio Television Handbook, Americans owned about 215 million.[8] Many U.S. households have more than one set, and nearly a third of these stopped working in the last five years.[9] The Japanese magazine *Asia Pacific Perspectives* estimates "that 150 million new television sets are sold each year worldwide, a number on par with annual sales of computers."[10]

Another important factor to consider when assessing the volume of e-waste is the rapid rate at which high-tech electronics have proliferated. For example, between 1992 and 2002, U.S. sales of consumer electronics, including PCs, quadrupled.[11] During roughly the same period of time—between 1991 and 2003—cell phone subscribers grew from 15 million to 740 million in the countries that make up the Organization for Economic Cooperation and Development (which includes the United States, Canada, Mexico, Japan, South Korea, Australia, New Zealand, EU countries, Norway, Sweden, Iceland, Turkey, and several eastern European countries).[12] In the United States sales of newer high-tech electronics, like DVD players, have grown similarly, rising from about one million in 1997 when they were introduced to twenty million in 2002.[13]

MOORE'S LAW AND E-WASTE
Combine this exploding quantity of electronics with the equipment's short life span and its plummeting resale value and you get what Lauren Roman, then vice president of marketing for United Recycling Industries, described as a "skyrocketing rate of obsolescence" at the 2002 International Symposium on Electronics and the Environment.[14] Thanks to what's become known as Moore's Law—that semiconductor power would double every eighteen months to two years, named for Gordon Moore,

cofounder of Intel who first made his observation about the exponential growth in the number of transistors per integrated circuit in 1965[15]—the average life span of a computer is about three years. The electronics industry estimates a cell phone's life span to be only two years, while a TV may last as long as thirteen or fifteen years.

While televisions last longer than any other consumer electronics, the electronics recycling industry considers TVs to be among the most challenging items to recycle, given the number of them that exist, their weight, the fragility and toxicity of CRTs, and their relatively low scrap value. Recycling efficiency depends to a large extent on having a quantity of relatively homogenous items from which to recover material that will become the feedstock for new products. Discarded TVs, unfortunately, are more variable than other electronics, running the gamut from your grandma's old veneer cabinet console to your neighbor's fourteen-inch portable.

Over the past ten years, about twenty million color TVs have been sold each year in the United States. With mass production of high-definition television on the horizon, the number of TVs that could soon be rendered obsolete is enormous. When and if broadcasters move to the digital format, as they are being pushed to do by the Federal Communications Commission and Congress, your conventional TV will no longer function as is. The current price for digital TVs remains high, but sales have been increasing annually—78 percent in 2004 alone.[16]

In 2004 the sales of CRTs (cathode ray tubes) and LCDs (flat-panel display screens) were about evenly matched, but it's expected that by 2008 CRTs will amount to no more than 20 percent of monitors and screens sold. As people discard older TVs and computer monitors, more CRTs will be entering the waste stream, thus increasing the need for safe and environmentally sound disposal systems.

As technology continues to evolve, and the system of production that keeps costs relatively low persists, any incentive that may exist for most consumers to repair or otherwise extend the life of high-tech electronics disappears. Unless this equation changes, we will continue to acquire newer and newer models, tossing more out as we go along. "One of the biggest problems we have is the throwaway mentality," said Don Cressin of the National Electronics Service Dealers Association in 2002.[17] Cressin has seen the number of service dealers in the United States shrink by two-thirds since 1980. High-tech electronics are resource intensive to produce, lose value quickly once in use[18]—thanks to ever-improving technology—and are expensive to dispose of. Sego Jackson, principal solid waste planner for Snohomish County, Washington, calls this cycle "dysfunctional."[19] Making it functional, with hardware and software that extend the life of high-tech equipment and make it easier to recycle, could help solve some of the industry's knottiest environmental problems.

HOW MUCH OF AN E-WASTE TSUNAMI IS THERE?

The United Nations Environment Programme says that "e-waste represents the biggest and fastest growing manufacturing waste."[20] But the volume of e-waste and quantities of the specific electronics that comprise it are difficult numbers to get a grip on—and the history of tracking this information is short. The original equipment manufacturers —the companies that make high-tech electronics—know how many of their products they've sold. Some rates of disposal have been estimated by comparing these numbers to the expected useful life span of a particular piece of equipment. Others have been calculated by piecing together volumes or units of discarded equipment.

In the United States, the International Association of Electronics Recyclers, the National Safety Council, some institutions, and organizations like the Silicon Valley Toxics Coalition and others that comprise the Computer TakeBack Campaign, along with individual states and regional recycling coalitions, have begun collecting and analyzing such data, as have a number of other countries. But as of the end of 2004 there was no central database for this information and no really precise numbers for the amount of e-waste being recycled. Nor does there currently exist any uniform way of accounting for e-waste from country to country. Descriptions abound, however. One reports that the United States discards enough e-waste annually to cover a football field a mile high.[21]

We know that an enormous amount of used high-tech electronics is simply trashed domestically—ending up in municipal landfills and, in some places, incinerators. Some equipment is sent to domestic recyclers, some equipment is reused as is; some is refurbished and given or sold to subsequent users, while some is dismantled for usable parts. Some e-waste (whole, intact equipment or parts) is exported—primarily to Asia and other developing countries—for cheap recycling. When I asked in late 2004 whether anyone was initiating systematic collection of such information, a U.S. Environmental Protection Agency official commented, "I don't think this is something anybody on the planet can do for you."

Discarded electronics, we do know, are the fastest growing component of municipal trash both in the United States and in Europe. The International Association of Electronics Recyclers (IAER), which has done the most detailed surveys of U.S. e-waste to date, expects that somewhere in the neighborhood of "3 billion units of consumer electronics will become potential scrap between 2003 and 2010."[22] They expect 250 million computers to become obsolete between 2007 and 2008 alone[†] and that at least 200 million televisions—about 25 million a year—will be discarded between 2003 and 2010.[23] The Silicon Valley Toxics Coalition has parsed this flow of U.S. high-tech trash further, estimating that by 2006 over 163,000 computers and TVs will become obsolete every day—a bundle of e-waste likely to weigh in at more than 3,500 tons.[24] As for cell phones, the high-tech item with the fastest turnover, the EPA anticipates that by the end of 2005 discarded cell phones will have created approximately 65,000 tons of waste.[25]

So where has all the e-waste gone?

The EPA estimates that, on average, over 2 million of tons of e-waste find their way to U.S. landfills each year.[26] And high-tech trash has been accumulating steadily. In 1997 some 3.2 million tons of high-tech electronics waste were dumped in U.S. landfills, while in 1998 over twenty million computers were discarded in the United States—a volume the EPA estimated to be increasing by 3 to 5 percent a year.[27] According to the IAER, we're discarding some 100 million computers annually.[28] The National Safety Council estimates that 85 percent of the 63 million U.S.

† The IAER estimate coincides with the EPA's May 2002 estimate of over 250 million computers expected to be retired from use by 2007.

computers discarded in 2002 probably ended up in landfills and that only about 10 percent of all discarded electronics were recycled.[29] The estimate for the percentage of e-waste being recycled in Europe (prior to the WEEE directive) is about the same as it is in the United States, with some 90 percent thought to be ending up in EU landfills or incinerators.[30]

A bulletin released by the EPA at the end of December 2004 reported that the U.S. government alone currently disposes of approximately 10,000 computers every week—a number that doesn't include computers used by the military or the U.S. Postal Service. "A significant portion" of these computers, says the EPA, is sent to landfills, stored, or is exported for cheap, often environmentally hazardous recycling.[31] When I asked what "a significant portion" might amount to, officials said they had no hard numbers. What has the U.S. government been doing with its old high-tech equipment that isn't tossed or squirreled away? Some has been auctioned off by the General Services Administration and some has been acquired by recyclers or by nonprofit organizations and schools. The federal government's first ongoing contracts with electronics recyclers were not instituted until the very end of 2004.[32]

▌▐ ▌▌▌ ▌

Despite the unique logistical challenges and environmental problems posed by e-waste, and the quantity that other writers have characterized as a tsunami and Brobdingnagian, the EPA has not in the past separated electronics out as a distinct category of refuse. The system has bundled the high-tech trash in with other discarded appliances, so-called durable goods—typically vehicles and household appliances, like refrigerators and washing machines—expected to last five years or more. In early 2005, however, the EPA announced that electronics would, in the future, be classified and counted as a discrete type of solid waste. This may seem an arcane technicality, but as observed by Wayne Rifer, an environmental consultant who specializes in e-waste issues, "What you measure, you manage."[33]

Across the Atlantic in the European Union, the e-waste has been piling up just about as prodigiously. But there, regulations mandating electronics recycling going into effect in 2005 and 2006—the WEEE directive (Waste from Electrical and Electronic Equipment)—are designed to change this status quo. Parts of Europe are already recycling electronics through national and regional collections systems, but the EU-wide mandate for electronics recycling will harmonize the process across the continent. Because more electronics recycling programs are in place in Europe, those countries have a better idea of what's being collected, but overall e-waste numbers are based on the same kind of estimates as in the United States.

The United Kingdom's estimated rate of electronics recycling has been no better than that of the United States, or that of the EU as a whole before regulations have gone into effect—about 10 percent. In December 2004 the BBC reported that 200 million electrical items per year, including about 2 million computers and 2 million televisions, were ending up in ordinary UK landfills—rather than in those created to handle hazardous waste—and that there could be some 6 billion items of high-tech trash currently buried across Great Britain. To dramatize the gargantuan size of Britain's e-waste problem, the Royal Society of Arts is building a sculpture nearly twenty-three feet tall that will weigh 6,600 pounds, dimensions that are supposed to represent the amount of e-waste one person disposes of in a lifetime.[34] This giant—a hideous cadaverous figure to judge from the sketch I saw—created to illustrate the monster of waste unleashed by high-tech will be called "WEEE Man." It will first tower over the South Bank of the Thames in London and will later tour Britain and France.

Add to Britain's e-waste the rest of the electronics discarded in the European Union each year and you end up with over 6 million tons—a volume that is growing 3 to 5 percent a year.[35] In Japan where home appliance recycling became mandatory in 2001, followed by a 2003 law that requires recycling of PCs owned by individuals, Sony reports recycling rates of 86 percent for its televisions and 70 percent for its desktop PCs. Altogether, in 2004 Sony says it recycled over 15,000 tons of equipment.[36]

WHAT'S BEING RECYCLED?

Most of the scant 10 percent of discarded computers and related high-tech electronics and peripherals now recycled in the United States come from corporations, institutions, or governments that purchase large quantities of equipment—or from programs designed to collect specific components, such as printer and toner cartridges or batteries. A 1999 National Safety Council study found that over 80 percent of recycled electronics in the United States came from the original manufacturers of these products and from companies with over five hundred employees, as did over 75 percent of the equipment received by companies and organizations that refurbish and resell used electronics.[37] According to an IAER study, little had changed in this regard four years later. In the United States the IAER found that government, schools, and consumers all recycled about the same amount of their used high-tech electronics—proportionally less than half the amount that manufacturers and industry did.[38]

Because most of the electronics now recycled come from large businesses and institutions, what we know about the used and obsolete high-tech electronics coming from small businesses and homes is fairly sketchy. One study reports "that 70 to 80 percent of old home and office computers in the United States are stockpiled before any other option is considered,"[39] while a survey released in January 2005 found that over half of all U.S. households have working electronics that are no longer being used.[40] To get a picture of how much high-tech trash there may be around the United States, I gathered estimates from a few states. These numbers reflect a combination of e-waste reports gathered from small businesses, individual households, and in some cases from local government offices. What impressed me is that even small states measured their e-waste in hundreds of thousands of units that weigh thousands of tons.

In 2001, Oregon (where I live)—with a population of just under 3.5 million that year—discarded about half a million computers, or 10,000 tons' worth.[41] That same year in California (population about 35 million), where 6,000 computers are estimated to become obsolete each day, the state's waste management board estimated there were some 6 million old computer monitors and TVs gathering dust.[42] Meanwhile in the Northeast, Maine's nearly 1.3 million residents have been abandoning some 100,000 computers and TVs each year, while Massachusetts—one of the few states that has any curbside recycling for electronics—estimated that the state was generating about 75,000 tons of obsolete electronics a year in 2003—a volume expected to grow to 300,000 tons per year by 2005.[43] And in Canada, Environment

Canada reports that in 1999 PCs were responsible for 13.5 tonnes (about 29,700 pounds) of lead, 2 tonnes (about 4,400 pounds) of mercury, and 0.5 tonne (about 1,100 pounds) of cadmium.[44] The annual estimate for Canadian e-waste disposal in 2005 is somewhere on the high side of 70,000 metric tons.

A number of high-tech manufacturers have launched electronics recycling programs and held collection events, often in cooperation with local governments, retailers, and nonprofits. These usually short-term events have collected the majority of electronics destined for recycling in the United States.[45] In the summer of 2004 a collection program run by Office Depot and Hewlett-Packard at all 850 Office Depot stores collected about 10.5 million pounds of equipment over the course of six weeks. Hewlett-Packard's Web site says that the company reuses or recycles "over 3.5 million pounds of product" in the European Union and United States each month. But in 2000 IBM—the first manufacturer to have a fee-based computer recycling program—recovered less than 1,000 computers through a similar program, less than 0.03 percent of that year's annual sales.[46] Best Buy's used electronic collection events, held seasonally since 2001, have collected over 2 million pounds of equipment, over 750,000 pounds in 2004 alone. Those who like to shop for bargains that benefit charities might like to know that in 2004 people clearing out their homes and offices donated over 23 million pounds of used electronics to Goodwill.[47]

"Pressure from residential communities for recycling is intense," says Kate Krebs, executive director of the four-hundred-member National Recycling Coalition. In the fall of 2004 nearly seven hundred counties and cities in the United States were collecting used electronics in some fashion.[48] But as of January 2005 only a handful of states had legislatively mandated e-waste regulations or ongoing recycling programs either under way or under consideration, and most existing programs were in their infancy. And while people participating in e-waste recycling and high-tech reuse programs are enthusiastic unloaders of equipment, overall participation remains small. A 2004 cell phone collection event in New York's Westchester County, which has a population of over 900,000, collected only thirty-two phones.[49]

Possibly contributing to the current sporadic participation in electronics recycling is the fact that no two companies or states do exactly the same thing, creating confusion for customers and frustration for industry. While large-scale purchasers are beginning to return used high-tech equipment to manufacturers on a regular basis, none of the existing take-back programs really has the capacity to capture the backlog of e-waste generated by households and small businesses.

"It should be as easy to recycle a computer as it is to buy one," says Sego Jackson.[50] But it's not. The information is out there, but you have to want to look for it and be willing to sort out the options. In March 2004, for example, Dell's Web site featured a recycling special: for $5 ($10 off the regular price) Dell would pick up and deliver to their recycler up to fifty pounds of any manufacturer's computer equipment. This was a much better deal than Apple's program, which cost $30 dollars and required drop-off at UPS. On the same day Hewlett-Packard's recycling prices ranged from $13 to $34 per box, prices that the company's Web site said were "subject to change without notice."

In general, if you live in the United States, unless you're a resident of one of the Massachusetts communities that have added electronics to their curbside recycling programs, it takes a lot of effort to figure out how to recycle even something as small as a computer printer ink cartridge. For example, it took about half a dozen phone calls and numerous Internet inquiries—some of which yielded incorrect information—to find out where I can take my Canon printer cartridges for reliable recycling.

WHAT'S IN YOUR CLOSET?

To find out what my home state is thinking about in terms of e-waste policy, I sat in on a few meetings of Oregon's Electronic Product Stewardship Advisory Committee. At one meeting I heard a representative of the Electronics Industry Alliance wonder about the extent of the state's e-waste and a local government representative talk about initiating a survey to determine the region's volume of e-waste. That survey was going to be conducted by contacting area households, so I decided to see what a survey of my own would yield.

My sampling was limited to people of my acquaintance rather than a cross section of the general public. My survey participants, I will admit, are environmentally well educated and probably fussier in their consumer habits than average, but the results are in keeping with trends identified by the electronics industry and others and are indicative of the quantities of non-biodegradable trash piling up all over the world.

I conducted the survey in June 2004 and got a response rate of about 50 percent. Altogether my twenty-six respondents had ninety-two pieces of stored, obsolete, or no longer used high-tech electronics, plus a substantial number of unused floppy and zip disks. The breakdown went like this:

Fifty-eight percent of respondents had cell phones (one respondent had also recycled three additional cell phones in 2003).

Half had laptop/notebook computers and monitors (one respondent had recycled four additional laptops in 2003).

Fifty-four percent had zip and floppy disks (and half of these respondents said they had "lots" of disks).

Forty-two percent had cordless phones.

Twenty-seven percent had printers.

Fifteen percent had televisions.

Twelve percent had fax machines.

Eight percent had VCRs, digital or video cameras, PDAs, and power cables.

Four percent had old CD or video game players in storage but all respondents who had used equipment had at least one other component—including keyboards, modems, speakers, towers, or zip drives—in storage.

Nineteen percent of the respondents had nothing in storage (so it's actually twenty-one people, rather than twenty-six, who had ninety-two pieces of equipment plus piles of disks filling up their storage space).

In addition to asking what people might have in storage, the questionnaire also asked, "Why is this stuff in your closet?"

"I already know it shouldn't be tossed in the trash." 50 percent yes. This response indicated that my survey participants were better educated on the subject than other segments of the population. A survey undertaken in Florida revealed that over 80 percent of consumers there didn't know that computers could be recycled.[51] And a survey released in January 2005 found that only 15 percent of Americans know that electronic equipment can be recycled, and that most people don't know what their local recycling options may be.[52]

"I figure someone else may be able to use it someday." 27 percent yes.

"I kept it as backup in case my new equipment didn't work and I've procrastinated throwing it away." 27 percent yes.

"I may use it myself someday." 23 percent yes.

"I am a packrat in other aspects of my life as well." 19 percent yes.

"I just couldn't stand the thought of throwing away something that cost so much that I used for such a short time." 15 percent yes.

"Nostalgia." 8 percent yes (for example, "I do have the 1983 Mac 512K that I wrote my first manuscript on stuffed away in a closet—for sentimental purposes. Someday I hope to covert it to a fish tank").

Others variously mentioned that equipment was on the way to Goodwill, waiting for a good recycling option, or asked me what I recommended. One friend told of her attempts to find a recipient for her station wagon full of old electronic equipment: "These old computers were perfectly useable, they just weren't . . . well, powerful enough. The problem was that they all have memories in them, and the memories were ours. We weren't savvy enough to clean the memories, so we didn't want to just give the things to someone who might strip them of information that we didn't want shared. But we had heard rumors that somebody was sending old computers to Belgrade or somewhere, and we were happy to have them used. We called around, asked around, drove around, and finally found somebody who would have been happy to charge us to take the things. So we paid a little bit, and I presume he stripped them and sold the parts somehow. I don't know. Computers should have a switch that wipes them clean. Then I bet people would recycle them more readily."

There is software (some of which can be downloaded free of charge) that does wipe a computer's hard drive clean, but this information does not come with a computer. The absence of such information is clearly making many people hesitant to move their old, unused equipment out of storage—one reason that the dilemma of ridding old equipment of personal, professional, or proprietary data persists.

It's no wonder that disposing of a used computer is a conundrum. Like nearly all consumer products, most high-tech electronics sold in the United States come without any information whatsoever about environmentally sound disposal and reuse options. A few have cryptic instructions, like the pictograph on the instruction manual that came with the Siemens cell phone I bought in 2004. It shows a cell phone poised over a trash bin, on top of which is printed the universal "no" symbol, the red circle with diagonal line. But almost without exception, getting rid of e-waste in a responsible way is a time-consuming endeavor even for conscientious consumers. One consumer survey conducted for the Wireless Foundation and the Cellular Communications and Internet Association found that 60 percent of those surveyed were given no information on what to do with an old cell phone when they purchased a new one and that 60 percent of those same consumers had an old cell phone at home.[53]

The disposable consumer society that we have become is a very hard habit to break. A new start—whether it's a new sheet of paper, a new career, or a brand new computer—is part of the American dream. But the situation is beginning to change with respect to electronic equipment, thanks largely to what's happening in Europe, as unpopular as that may be with many U.S. businesses and policy makers. The Electronics Industry Alliance writes on their Web site that "the U.S. high-tech industry is the only industry whose products become smaller, cheaper, better, faster and more environmentally friendly year after year. EIA and members . . . proactively address environmental concerns and actively work to reduce the environmental impact of electronic products and manufacturing processes where technically feasible through policy and advocacy work and voluntary industry design for environment tools." Yet the same policy statement also says, "Despite EIA's voluntary environmental initiatives, a growing number of nations and states are considering legislation and regulations, which would severely impact the ability of the electronics industry to ship and sell its products globally."[54]

"In the last eight to ten years Europe started to look at making manufacturers play a role in their products' end of life," Heather Bowman, then director of EIA's environmental affairs, told me in February 2004. Turning to this side of the Atlantic, she continued, "As responsible companies, U.S. manufacturers are involved in pilot projects and voluntary programs to increase the use of recycled materials and for shared responsibility at end of product life. Today we want to take care of the entire program at a national level and we need legislation to make a level playing field."

"In a competitive business, it's a very difficult issue," Bowman said. "Everyone has to have the same rules." Without action at the federal level, manufacturers are looking at the potential of fifty different sets of regulations within the United States, plus those coming from Europe and elsewhere. (Industry thinking on the issue has evolved quickly. In 2002, only two years before my conversation with Bowman, she addressed the International Symposium on Electronics and the Environment, saying, "What's needed is education not legislation."[55])

So it seems that even from an industry perspective the gospel of letting the market solve the problem all by itself may not work with e-waste. Without some regulation to systematize collection of e-waste and ensure a steady stream of recoverable materials, it may not be possible to turn electronics recycling into a viable business. And without regulation it probably won't be possible to ensure that electronics recycling takes place under environmentally and socially responsible conditions. "A while ago industry was taking an NRA-like approach—an over my dead body approach—but that's changed enormously with the EU directives," Dave McCurdy, president of the EIA and former U.S. representative from Oklahoma, told the audience at the E-Scrap 2004 conference.

"POISON PCS AND TOXIC TVS"

The details of e-waste policy may lack drama on their own, but they have been the subject of intense debate on all sides of the Atlantic and the

Pacific. The policies that will really make electronics recycling successful represent a radical departure from traditional ways of doing business. That sounds rather grandiose, but e-waste is a pretty quirky issue. It has brought together activists from the Philippines, the Netherlands, Hong Kong, India, London, California, Texas, and Seattle, with academics, engineers, and researchers from China, Sweden, New York, Tokyo, and beyond, and with executives from the world's leading high-tech manufacturers and mining companies. The process of hammering out e-waste policy might make even the most dedicated lawmaker long for the lunch break. But these are the small steps that will—with some luck—begin to shift how we account for the true costs of high-tech products and prompt the design of more environmentally friendly products.

It's safe to say that none of the changes Dave McCurdy alluded to would be happening if not for the agitation of environmental and consumer advocates. In the United States, at the heart of these efforts are the Clean Computer Campaign initiated by the Silicon Valley Toxics Coalition, and the Computer TakeBack Campaign, launched in 2001 by a coalition of nonprofits that includes the GrassRoots Recycling Network, Silicon Valley Toxics Coalition, Texas Campaign for the Environment, Clean Water Action, and Materials for the Future Foundation.

In seeking to influence the high-tech electronics industry, the environmental advocates have taken on a formidable task. The EIA has well over a thousand members and represents an industry with annual revenue of somewhere around four hundred billion dollars. But the environmental advocates watchdogging the industry also have a global reach, and like industry they're using this strategy to their advantage. Silicon Valley Toxics Coalition and their Computer TakeBack Campaign colleagues in the United States work closely with the Basel Action Network, Clean Production Action, and Greenpeace International. There are policy issues specific to each individual country, but when it comes to pushing high-tech companies to make "cleaner" products, this might be a rare instance when environmental advocates will say that globalization can do some good.

In the late 1990s the Clean Computer Campaign began releasing its annual report card, rating high-tech manufacturers on the use of hazardous materials in computers, the ability to upgrade existing machines, and to return old computers to the producer for safe recycling and reuse. These reports and an assessment of e-waste called *Poison PCs and Toxic TVs*, first published by the Silicon Valley Toxics Coalition in 2001 and updated several times since, have garnered major news media attention. The companies whose products the reports target have also taken serious notice. After the 2002 release of *Exporting Harm* by the Basel Action Network—a powerful exposé about the export of e-waste to Asia[56]—it became impossible for U.S.-based high tech not to give high priority to environmental issues, especially since these reports coincided with enactment of e-waste policy in Europe. These groups have also pushed the industry on its use of prison labor for recycling. After release of a report on the subject in the summer of 2003, and some embarrassing public demonstrations by protestors wearing prison uniforms and calling themselves a "high-tech chain gang," Dell quickly changed its policies.

E-waste raises issues that affect an interesting cross section of business, government, and consumer interests: water quality, waste management, liability, fair labor practices, data security, and regulation of a multibillion-dollar industry whose products most of us can't imagine living without. The Clean Computer and Computer TakeBack campaigns' advocates have made the most of this, by—as one industry observer put it—"keeping the pressure on, not only with its scathing reports but by pressuring local governments to pass resolutions favoring new state laws."[57]

The sheer bulk of used electronics has turned local governments who are interested in shifting some of the burden of e-waste off taxpayers into the Computer TakeBack Campaign's "biggest allies," David Wood told me in early 2004 when he was executive director of the GrassRoots Recycling Network. "The cost of handling and processing e-waste could be enormous and prohibitive," said Wood. According to the Computer TakeBack Campaign, which did its assessment in 2003 and 2004, the cost of collecting and processing e-waste over the next ten years could exceed ten billion dollars.

Not surprisingly, this has drawn considerable attention from legislators and other local officials all across the country. Solid waste is not glamorous, but getting rid of it costs money and so it's a staple on every state and local budget. If not handled properly the e-waste burden goes far beyond the price of setting up a recycling system. It has the potential to cost community health and environment dearly. Say "lead" or "mercury" or "dioxin" to a local policy maker and you will have her attention. And the state e-waste bills passed thus far have been aimed at keeping lead and other heavy metals out of local landfills and at curtailing the use of toxic chemicals in high-tech electronics. "The states are the only places we've seen any progressive policy on toxics in the last decade," remarked Wood.

Since its inception the Computer TakeBack Campaign has worked with communities to craft legislation to control the hazards of e-waste and with manufacturers and retailers on used electronics collection events. Fundamental to the campaign is the concept of producer responsibility. As Ted Smith, founder of Silicon Valley Toxics Coalition has said, "I think a lot of our perspective has been shaped by contacts with some of our key allies in Europe that have been focusing on product policy for the last several years—again, way before anybody in the U.S. really started thinking much about it."[58]

"We first started pushing the idea of producer responsibility for e-waste in mid-2000," Wood told me. "In the early stages of the campaign, we heard from office managers and property managers who had old equipment in storage. Initially the campaign put pressure on Dell—the leader in terms of PC sales—because Dell's business model presents real opportunities for take-back. This proved to be a very wise choice, because when Dell moves, the rest of the industry moves to stay competitive. The campaign turned an industry laggard, in terms of take-back, to an industry leader."

"The main programmatic things we're working on now are national computer take-back and materials [improvements] and trying to import what's happening in Europe, Japan, and elsewhere," Smith told me in March 2004. "We're doing this on a state-by-state basis, promoting legislation at the state level, as we're never going to get anything out of Washington," said Smith. "With enough states weighing in, the feds will have to step in and harmonize."

So if the high-tech market leaders like Dell and Hewlett-Packard—and other major manufacturers—have their own take-back programs in place, what's the problem? What more could the environmental advocates want and what's the remaining controversy about? Many environmental advocates and some solid-waste professionals find manufacturers' existing programs insufficient—in scope and convenience—but at the heart of this debate is the pesky concept of producer responsibility.

PRODUCER RESPONSIBILITY VS. PRODUCT STEWARDSHIP

E-waste is one of the most complex, bulky, and nonbiodegradable forms of trash ever. This makes dealing with it in an environmentally sound way very expensive. In the United States, garbage—both trash and recycling—is managed and financed by local governments and taxpayers. But the high price of processing e-waste and its environmental liabilities are prompting local governments, some institutions and businesses, as well as consumers and environmental advocates, to question this status quo. Fundamental to this discussion is the idea that end-of-life product costs—impacts and responsibilities traditionally borne by the community and environment—should be shifted to or shared by the manufacturer.

"Companies that manufacture products should retain physical or fiscal responsibility for those products when they become waste. If you connect the design process with the cost of waste, you create a great impetus for designing products that contain less toxics and create less waste," Joanna Underwood, president of INFORM, a New York nonprofit that studies the environmental impact of business practices, told me in March 2004.

This concept is known in most of the world as extended producer responsibility (EPR), but is often referred to in the United States as product stewardship, which EPR purists are quick to point out is not equivalent to EPR. The concept is new to Americans. "Europe is way ahead in this, and it's coming very late to the United States," said an American expert in electronics recycling policy who asked not to be named. So alien to the traditional American mode of business is the idea that the manufacturer should play an active role in the disposal and recycling of its products, writes journalist Mark Shapiro, that the concept "has been received in this country like a message from another planet."[59]

In the early 1990s, the idea of extended producer responsibility began to gain traction in Europe as a way of dealing with waste. EPR "was first introduced in a report to the Swedish Ministry of the Environment in 1990," explains Iza Kruszewska of Greenpeace International and Clean Production Action.[60] The idea behind this was "to find environmentally sound solutions as far as possible ahead of the waste stage," which means thinking about how products will be disposed of when they are being manufactured, not just after they become defunct.[61] In 1991 the European Union introduced this approach for some individual product waste streams, which prompted a number of countries to involve manufacturers in the waste-collection process. In pure EPR the customer pays nothing extra for recycling, although presumably those costs may be folded into the price of new products. But forcing manufacturers to physically take possession of used product is key, as that's where the incentive to reduce overall materials and toxics comes into play.

Germany's legislatively mandated system for recycling packaging materials—the German Packaging Ordinance—has been in place since the early 1990s. Called Green Dot, the program requires manufacturers to take back used packaging from consumers free of charge and to send it for recycling or to help finance a system that collects packaging from individual households. Green Dot reduced packaging in Germany by 14 percent between 1991 and 1995, years in which packaging in the United States increased by 13 percent.[62]

The Green Dot program and others like it in Europe—for batteries, carpet, tires, and defunct vehicles, for example—along with a proliferation of ecolabels in northern Europe, set the stage for what became the European Union's RoHS and WEEE directives for reducing toxics in electronics and

for handling e-waste. Electronics manufacturers, nearly all of whom operate internationally, initially lobbied against these regulations, arguing that
they could impose trade barriers, have significant adverse financial impacts,
and could potentially encourage the use of methods and materials more
environmentally harmful than those already employed.[63]

Because of WEEE and RoHS, and Europe's precautionary approach
to chemicals used in consumer products, manufacturers doing business
there have been pushed to move substantially ahead of the United States
on policy regarding waste and toxics reduction and the recycling of obsolete products. I asked the electronics product stewardship expert who
asked not to be identified why might this be so. "U.S. industry has so much
more clout relative to government than they do in Europe. It has a culture
of not being regulated and resisting regulation. It's kind of the last bastion
of free-market capitalism," came the answer.

Another obstacle for the implementation of producer responsibility
policies in the United States, electronics recycling policy pros told me, is
that it's very unusual for American industries to sit down with their com-
petitors and discuss an issue that affects a company's business model or
product design. This is not peculiar to high tech, but the industry's com-
petitiveness highlights this difficulty. U.S. companies are not accustomed
to adapting their business models to conform to another's plan, to accommodate government regulation, or to bow to consumer concern over
issues of materials or design, let alone product disposal.

But pressure on U.S. electronics manufacturers to reform is coming from
groups not often cast as environmental activists: large-group-purchasing
organizations. The Western States Contracting Alliance, the Society of
College and University Professionals, and organizations that purchase
equipment for the health-care industry are among those working to
include environmental criteria in new purchasing agreements—criteria
like take-back, hazardous materials reductions, and recycling that bars
export of e-waste. A number of these groups and local governments
across the United States—some working as regional coalitions that

include nonprofits—are incorporating these requirements into environ-mentally preferable purchasing programs.‡ Among the motivations for these programs are cost savings—achieved by negotiating take-back as part of the overall service of a purchasing agreement—and concern over the liabilities posed by improper disposal of e-waste.

To find out what the federal government might be doing in this arena, I asked a couple of EPA officials if current federal government contracts with suppliers of high-tech equipment might include any manufacturer take-back provisions. "Some contracts might. But," said the officials who asked not to be named, "it wouldn't make that much of a difference, as ultimately the equipment would all end up in the same place: either with a recycler or being refurbished for reuse." Admittedly, the individuals I spoke with were not in a position to set policy, but their answer seems to confirm what solid waste manager Sego Jackson has said, "Americans are not yet literate in product stewardship."[64]

SO WHERE DOES THE HIGH-TECH TRASH GO?

If only 10 percent of U.S. e-waste is being recycled, where does the rest of it go? Where should it go? Why is it still so easy to toss high-tech equipment in the bin? And what are some other countries doing to tackle the problem?

Despite high-tech electronics' toxic contents, the United States has no national legislation regulating e-waste disposal and no national system for electronics recycling. This is in contrast to over a dozen other countries, including the Netherlands, Norway, Sweden, Belgium, Switzerland, Japan, Korea, Taiwan, and—as of 2006—those of the entire European Union. Several pieces of federal e-waste legislation have been introduced in both the U.S. House and the Senate, but none as drafted would really take care of the problem.

After three years of work to find a national solution through a volun-tary effort called the National Electronic Product Stewardship Initiative,

---

‡ Such efforts exist in the Northeast, Mid-Atlantic, Southeast, and elsewhere, including one called the Western Electronics Product Stewardship Initiative (WEPSI), which has been organized by representatives of federal, state, and local agencies and nonprofits from eight western states: Alaska, Arizona, California, Hawaii, Idaho, Nevada, Oregon, and Washington.

the process ended with a number of recommendations, but with no final plan. The result is that in the United States e-waste is being handled— where and if it's being handled—under what industry observers are fond of calling a "patchwork" of programs. While electronics recycling programs are growing, provisions for dealing with e-waste continue to fall far short of the rate and volume at which it's produced.[65]

So how does a country begin to regulate e-waste? First, you have to decide what kind of trash it is. Clearly, e-waste is solid waste; on that everyone can agree. But what kind of solid waste? Is e-waste hazardous, and if so, how should it be treated? If the market for high-tech electronics is global when the products are sold, is it correspondingly global when the products are discarded?

Whether high-tech electronics containing lead, chromium, other heavy metals, and toxics-infused plastics will be disposed of in ways that will allow toxins to seep into groundwater, the food web, and the air depends at this point on some pretty technical and wonky policy details. Among other things, these details determine policies that say whether or not it's permissible for rich countries like the United States to ship hazardous waste to poorer countries. These policies also determine whether Americans will change their current approach to disposing of consumer products and regulating the materials those products contain—and if so, how.

In the United States, when high-tech electronics are disposed of, they're officially categorized as solid waste. Because so many of their components and materials are hazardous, they generally fall under the jurisdiction of the Resource Conservation and Recovery Act (RCRA), the hazardous waste program administered by the EPA. However, waste disposed of by households and by small businesses—anyone who discards less than a hundred kilograms (about 220 pounds) of such hazardous solid waste a month—is exempt from federal RCRA regulations, unless local regulations say otherwise.

This small-quantity exemption makes it all too easy for many to dispose of e-waste in environmentally harmful ways. Because there's no national regulation regarding e-waste, and because trash disposal is regulated locally in the United States, unless your state or local community expressly prohibits dumping specific high-tech electronic components (such as CRTs) or the materials they contain (lead or mercury, for example)—and unless you're dumping over 220 pounds of e-waste a month, which would be a violation of federal law—it's perfectly legal to toss this waste in with the rest of your garbage. Even if some materials found in high-tech appliances are classified as hazardous and aren't accepted by your local landfill or garbage hauler, proper disposal now depends on voluntary compliance. Curbside recycling bins are given the once-over before they are pitched into the truck, but no one looks through your trash on its way to the dump.

Local waste haulers I spoke to said they try to do some triage when they see bulky pieces of electronics in the trash. When they can, the haulers or folks who manage the dump divert these items from the landfill stream. At the moment, however, education and conscience are often the only safeguards against putting modest quantities of e-waste into the bin. Given the proliferation of such items, this means significant amounts of high-tech trash continue to be disposed of in ways harmful to the environment.

‖|‖|‖ ‖

Then there are the complex regulations developed by the U.S. government—and other economically developed countries—that specify how items destined for recycling may be considered "products" or a "commodity" rather than waste. These rules enable the United States and other countries to exclude certain e-waste items from official classification as hazardous when they are exported for materials recovery—even if some of those individual substances are hazardous. Under WEEE e-waste can be exported, but only if the country exporting it can prove that the material will be treated under conditions equivalent to those where the waste originates.

In the United States, material destined for recycling or reuse may be classified as a product rather than as "waste" even if it contains hazardous materials—as do computer monitors and circuit boards—as long as the items remain intact and are not broken up. This policy is intended to encourage reuse and recycling by allowing these items to be sent outside the country to facilities that need or can handle them. But since the United States lacks any official mechanism to track the export of discarded high-tech electronics or how they are subsequently disposed of, the policy also creates a loophole that allows the export of copious amounts of e-waste.

While these regulations may make it less cumbersome to recycle some high-tech electronics components, it does not address the issue of *how* or *where* these products should be disposed of. For instance, are there any official criteria for what constitutes recycling? If CRTs, for example, go to a glass-to-glass recycling plant in Herculaneum, Missouri, is this okay? What about such a facility in Brazil? How about if they go to a backyard recycling workshop in Guiyu, China? The U.S. rule also fails to address in any comprehensive way the hazardous materials CRTs contain, or who should shoulder the practical and financial responsibilities of e-waste disposal and/or recycling.

CRTs, because of their high lead content and the limited availability of leaded glass recycling, fit neatly into the confusing category of something that is both hazardous and recyclable, and in some cases reusable. Circuit boards, because they contain the most valuable recoverable materials in used electronics—precious metals—as well as lead and other toxic materials, also fit into the category of a discarded, potentially hazardous item that has commercial value.

Because CRTs are more fragile and their lead content greater, they are potentially more hazardous than circuit boards. But if not treated properly in disassembly and materials recovery, circuit boards can unleash a whole soup of toxics. As of this writing Australia is the only country that does not allow circuit boards to be exported for recycling. So it's easy to understand how state legislators and other local policy makers—who are on the front lines of dealing with e-waste—find coming to grips with high-tech trash and communicating that information to the public more than occasionally perplexing.

▌▐║▌▐ ▌

"We have yet to arrive at a consensus at the national level," Dave McCurdy of the Electronic Industries Alliance told the E-Scrap 2004 conference. "As a result of this we have multiple activities at the state level. We need national intervention at the federal level. It's time for Congress and the administration to step up." Without such action, cautioned McCurdy, electronics recycling will be inefficient and potentially costly, and it will be difficult to establish a much-needed "national market for recycled products" or to achieve a "level playing field for manufacturers and retailers." Thus far, despite various voluntary "e-cycling" programs accessible through the EPA's Web site, no electronics recycling program with any national consistency or authority has emerged in the United States.

Since 2001 most of the official American efforts to regulate high-tech trash on a national scale have been discussed as part of the National Electronics Product Stewardship Initiative (NEPSI). What is—or was—NEPSI? At a conference in 2002 I heard an EPA representative describe it as a "multistakeholder, joint, shared, common process," working toward a "voluntary, shared, common responsibility for end-of-life electronics." This would give anyone who has ever served on a committee or participated in a similar working group a sense of why the process may not have zoomed forward. The idea behind NEPSI was to create a national system to handle e-waste that would be acceptable to manufacturers, state governments, and environmental and consumer advocates, as well as to consumers, retailers, and local governments. This system would handle e-waste generated by households and small businesses rather than by large-quantity purchasers who, presumably, would have access to other avenues for dealing with their used high-tech equipment.

The NEPSI group included over forty stakeholders, among them manufacturers, state and local governance groups, recyclers, environmental and consumer advocacy organizations, retailers, and—for about two and a half years—the EPA. The full group met about a dozen times in person and by phone hundreds of times. One participant jokingly remarked that NEPSI was single-handedly responsible for keeping all conference calling systems in the United States in business. NEPSI came to the end of its run at the end of 2004—without having been able to reach a definitive conclusion. In early 2005, summing up her tenure with NEPSI, coordinator Cat Wilt put it this way: "At our first meeting in 2001 I found out I was one month pregnant. Last Saturday we celebrated my daughter's third birthday. I hope I don't have to continue marking milestones in my daughter's life by e-waste management issues."

Even though NEPSI ended without formulating how the United States would handle e-waste, it's worth cataloguing what its participants did agree on, as whatever happens nationally will likely incorporate NEPSI's recommendations. The group agreed on the scope of products the system would cover: all equipment with a CRT or flat-panel display screen of nine inches or larger, CPUs, laptop and notebook computers, consumer desktop devices like printers and fax machines, and the small peripherals that go along with all this equipment. This would, however, leave out cell phones, many PDAs, and similar small devices, as well as things like iPods and other MP3 players. The group also agreed on a base level of service; environmentally sound management guidelines; the need for market incentives to make the system work; a front-end financing system; and the need for national e-waste legislation. This last item is what prompted the EPA to withdraw from the NEPSI discussions in November 2003, as technically—although some policy analysts differ on the fine points of interpretation—the EPA is not supposed to engage in any activities that promote legislation. "The EPA withdrawal was a psychological blow to the group," says Wilt, but the issue the talks broke down over was financing.

A key issue in crafting an effective national program is collection. The transportation of used electronics between collection points and where the recycling actually takes place contributes significantly to the cost of handling e-waste. The stuff is generally heavy and bulky and often fragile, so it can't be tossed around like a sack of cans or bundle of newspapers, especially since it's when electronics are broken or smashed that their toxics begin to leach out. And unless there's sufficient volume, the business is not cost effective for those dismantling the equipment or for those recovering or recycling materials. In Europe and Japan, where populations are more uniformly distributed and distances are relatively smaller, transportation of e-waste poses less of a challenge than it does in the United States. Here, the country's size, varied geography, and contrasts between rural and urban areas make it difficult for a collection system that works for the Boston suburbs to work in Wyoming. As NEPSI participants recognized, any system would have to be adapted to suit local needs.

But Canada, whose geography is as sprawling and diverse as the United States' and which has a much smaller population, seems to be tackling the e-waste challenge with less distress than its neighbor to the south. The Canadian electronics industry has formed—and funds—EPS Canada, an electronics product stewardship organization whose members include over a dozen multinational electronics manufacturers. "The group was deliberately kept small and has created a single industry voice for e-waste in Canada," David Betts, executive director of EPS Canada, explained to the E-Scrap 2004 audience.

By the end of 2004, EPS Canada and all Canadian provincial leaders had reached an agreement on key e-waste issues: a collection program designed with "opt-out" provisions for a set scope of products; a visible, nationally consistent system of advance recycling fees that would be phased out over time; principles for enforcement and compliance; and cooperative management by industry and provincial governments. Meanwhile, individual provinces have been formulating e-waste regulations, and by early 2005 programs were already in place in Alberta and Ontario, with a handful of others expected to follow shortly.

The NEPSI stakeholders and EPS Canada—like their European counterparts—agreed that proper management of high-tech electronics' hazardous materials is of paramount importance. They also agreed on the need to certify electronics recyclers and to track and monitor the downstream processing of e-waste. The NEPSI group also concurred on the need for a nongovernmental, national coordinating entity—perhaps a nonprofit third party organization—to jump-start and possibly run the system like those up and running in Europe.

THE EUROPEAN INFLUENCE

The WEEE directive covers all electronics products sold in the European Union. Under WEEE, electronics manufacturers, no matter their home country, are responsible for taking back and recycling their products, either directly or through a third party. WEEE also makes it possible for consumers to return used appliances without charge, either to local collection centers or to manufacturers when purchasing new equipment, with the costs for historic waste divided between manufacturers according to current market share for that type of equipment and with some fees charged to consumers. And because virtually all high-tech manufacturers (including leading American high-tech companies) sell their products internationally, all of these companies will be participating in some fashion in take-back and recycling under the WEEE directive. By the end of 2006 the program aims to collect an average of four kilograms per person (about 8.8 pounds), with an additional goal of a 75 percent rate of collection for discarded CRTs.

To ensure that recycling—particularly of hazardous substances—will be done properly, WEEE requires manufacturers to provide recyclers with product materials lists, something that's not been done before. And under WEEE, used electronics may be exported for recycling or reuse, but only if the receiving country has environmental standards equal to those of the equipment's home country. If the reuse or recycling programs in the receiving country would not be acceptable in Paris or Copenhagen, under WEEE the export is prohibited.

Electronics recycling programs comparable to WEEE are already in effect in Sweden and other northern European countries. Scandinavia's program is run by El-Kretsen, a third party nonprofit organization made up of producers, retailers, and importers of electronic equipment. It offers consumers free collection and recycling of e-waste under a system run in coordination with local governments.

The system is financed by fees included in the sale of new products and by fees—determined by a producer's market share and type of equipment manufactured—that the producers pay into El-Kretsen. Large-quantity purchasers are charged special fees, and there are additional fees for specialized equipment, such as that used by hospitals. Used electronics can be returned via retailers (who are required to collect used electronics where they sell new equipment—an issue for U.S. retailers) and to municipal collection points, or they can be taken directly to the producer. All recycling is handled by certified recyclers, and El-Kretsen monitors recycling operations and tracks their performance for local and national environmental authorities. If a producer does not want to join El-Kretsen, it must show that it has a take-back and recycling system comparable to those of the directive. Over four hundred producers—international and Swedish—belong to El-Kretsen, which covers over 90 percent of all electronics sold in Sweden.

In Japan, which has an analogous electronics recycling system, the advance recycling fees collected on electronics go into funds managed by two different groups of manufacturers—those who do their own recycling and those who contract out for recycling. If the costs of recycling exceed the revenue brought in by the fees, the producers rather than consumers, government, or taxpayers absorb the costs.

"We're trying to build this into our business," Kevin Farnam, Hewlett-Packard's manager of corporate sustainability strategies, told me in 2004. "We see it as a service to our customers both large and small. Our position has been to work with the government to put in place the most efficient return system possible. It's shared responsibility, with the company responsible for a certain portion. Our responsibility is from the point where it can be colleted en masse and taken into the recycling system. We make these things so we have a good idea of how to take these things apart. We'll be leveraging off the existing system. In the EU we'll work with outside recyclers and vendors who've set up a consortia," said Farnam.

Even though the European e-waste collection systems include mechanisms to accommodate different manufacturers' financing needs, the NEPSI gang failed to agree on any. Those opposed to charging a fee, visible or hidden, called it a tax and feared it would hurt sales. Proponents of producer responsibility thought the manufacturer should bear the entire recycling fee, though in reality prices would probably reflect that cost. Still another camp—primarily manufacturers with existing take-back programs—was comfortable with charging fees for recycling and participating in that process, but wanted to retain autonomy and flexibility for themselves rather than have government set the program structure. Older companies like IBM with large amounts of historical waste, and with fewer individual consumers, were more concerned about recouping costs of collecting and recycling older products. There was also disagreement about who should pay for so-called orphaned waste—electronics manufactured by companies no longer in business.

The NEPSI group was also uncertain about the role of retailers—some of whom are also manufacturers in that they sell high-tech electronics under house brand names. An executive from Target who spoke at the 2005 EPA e-waste summit worried that responsibility for explaining recycling fees to customers would be left to the cashier who might be sixteen years old. In early 2005, online receipts from Staples—the office supply chain—came with a note saying, "Note to CA residents: You may have purchased some items that are subject to a recycling charge under the CA Electronic Waste Recycling Act. At this time, Staples has paid this fee for you."

In the spring of 2002, NEPSI agreed in principle to the idea of charging consumers a take-back fee to enable electronics manufacturers to recover and facilitate recycling of their products. The group then spent the next two years trying to hammer out exactly how such a fee would be levied. In the end, the NEPSI stakeholders agreed that the most workable solution would be some kind of hybrid model that would begin with an advance recovery fee and evolve into a hidden fee (called a partial cost internalization program). But they fell out on how to reconcile this plan with the manufacturers' varying business models and corporate cultures.

In early 2005, after the end of NEPSI discussions, federal legislation was introduced in both the House and the Senate. But neither bill truly addressed how a comprehensive national system of collecting e-waste for reuse and recycling would be financed. The bill introduced by Representatives Mike Thompson of California and Louise Slaughter of New York included a ten-dollar fee on new computers. This fee would fund an EPA program that would direct grant money to municipalities and organizations running electronics recycling programs; manufacturers who have existing recycling programs would be exempt from the fee—a list that would include virtually all major brands. The bill introduced by Senators Ron Wyden of Oregon and Jim Talent of Missouri would grant tax credits to companies and individuals recycling electronics and begin by covering only computers and televisions. But as introduced, the Senate bill had no details for how the recycling itself would be paid for.

At the electronics recycling summit held by the EPA in early 2005, I asked several participants—retailers, state officials, and recyclers—if it would make sense for the United States to adapt the European models instead of trying to reinvent this particular wheel. All shook their heads and said they thought that taking anything from Europe was highly unlikely, simply because it was European. But, as Boliden's Theo Lehner told a conference audience in 2003, "The U.S. is three years behind Europe in the legislative mechanism and can learn and benefit from our mistakes."[66]

WHAT THE STATES ARE DOING

"What's going on in the U.S. right now is like what went on in Europe ten years ago," said Robin Schneider, executive director of Texas Campaign for the Environment, at the E-Scrap 2004 conference.

Schneider was referring to what happened in Europe in the mid to late 1990s, when countries in Europe—Sweden, Austria, and the Netherlands among them—began passing legislation regulating disposal of e-waste. There was concern then, as there is now in the United States, that if individual countries—or states—enacted different requirements for handling e-waste it would cause difficulties for manufacturers selling to Europe (or the United States) as a whole. To homogenize the system, the European Union introduced the first iterations of the WEEE and RoHS directives in December 2000. Both directives cover a far broader scope of products than what any U.S. state legislation does—or what the proposed federal e-waste legislation would cover. Both WEEE and RoHS cover electronics of all kinds—virtually everything with a plug, from computers and televisions, to small devices like cell phones and MP3 players, as well as appliances like microwaves, hair dryers, electronic toys, refrigerators, and lamps.

The EU directives are not a panacea for e-waste. And advocates of producer responsibility, like Iza Kruszewska of Greenpeace International, don't consider the WEEE and RoHS directives ideal solutions. The directives aren't accompanied by a comprehensive ban on dumping electronics in municipal waste, and the details of exactly how to pay for transporting e-waste from collection points to recyclers in each jurisdiction are still being worked out. There are also numerous exemptions to the restrictions on hazardous materials—including lead in CRTs (thus far there is no alternative) and mercury lamps in flat panel display screens.

In the absence of a national policy or legislation in the United States, there has been an accelerating flurry of e-waste regulation bills at the state and local level. The escalating concern over the issue is remarkable. In the spring of 2002 ten states had about twenty or so electronics recycling bills under consideration.[67] By the 2004 and 2005 legislation sessions, over thirty states had introduced over four dozen e-waste bills, and a dozen states had passed some kind of legislation concerning e-waste.[68] And on the local level in 2003, about 250 ordinances covering e-waste were introduced, mostly in Massachusetts and California.[69]

Some states—notably California, Maine, Massachusetts, and Minnesota—have banned CRTs from their landfills, a ban that targets computer monitors and televisions. California has added to its ban a prohibition on disposing of flat-panel display screens or LCD and plasma screens in state landfills. Maine and California have passed legislation that will bar specific hazardous substances commonly found in electronics from products sold in state—these substances include mercury, cadmium, hexavalent chromium, lead, and certain brominated flame retardants. These bills essentially bring the European Union's RoHS directive to the United States.

Maine and Maryland have passed e-waste take-back bills that require manufacturers selling electronics in those states to have recycling programs. A number of other states have initiated used electronics recycling and collection programs, while others have passed bills funding e-waste studies or working groups. The state take-back programs run the gamut from pilot programs that establish ongoing and one-day collection events to those that establish regular curbside pickup. For example, Iowa began its electronics recycling program with one-day collection events that charged five dollars per item. In North Cook County, Illinois, over fifteen hundred vehicles came to deposit used electronics in two days of a drop-off program. A Pennsylvania drop-off day attracted a thousand vehicles loaded down with old electronics. A survey conducted by the Northeast Recycling Council in 2002 found that the average load per vehicle at used electronics drop-off events was 132 pounds.

In Massachusetts about 150 cities and towns have passed some sort of resolution supporting producer responsibility for products at the end of their useful lives,[70] and over 275 Massachusetts cities and towns collect electronics for recycling—a number of these at curbside. The state has nearly two hundred drop-off sites, which means that close to 95 percent of Massachusetts residents have convenient access to electronics recycling.[71]

Many e-waste collection events have been, and continue to be, conducted as part of cooperative agreements between retailers—including giants like Best Buy, Office Depot, and Staples—manufacturers, and local governments. Numerous manufacturers—including Hewlett-Packard, IBM, Sony, Sharp, Panasonic, Philips, Apple, and Dell—have electronics take-back programs for their own products and often for other manufacturers' equipment as well. Dell has an ongoing program with its home city of Austin, Texas, and has worked on pilot programs to collect e-waste from the New York City Department of Education and the Chicago public school system, among others. Lest this sound overly rosy, consider that a 2003 study by Snohomish County, Washington's "Take It Back Network," which tested electronics manufacturers' recycling programs found it took an average of two hours online to access recycling services and that several programs involved at least a week's wait for packing and shipping materials.

Why isn't there better coordination among these programs in the United States to make things less confusing for consumers, recyclers, local governments, and manufacturers? "If you can do it in the EU without erosion of market share and profits, why not do it here?" David Wood asked rhetorically when we spoke in 2004. "Electronics recycling will cost money; to spend money now so as to alleviate future environmental problems seems to me the more prudent approach. The burden," said Wood, "should not fall solely on local government and taxpayers."

"We see take-back as providing a service to the customer. Our biggest motivation is 'let's keep trying to make it better.' Let's make it the best in the industry. Dell works with audited recyclers and does not export," Dell's Bryant Hilton told me in 2004. "Here in the U.S. Dell will pick up any brand computer equipment from customers for a nominal fee. We can do this because Dell doesn't sell through retailers. We sell directly. Over half the business comes online and Dell has a big catalog program. This means that Dell is well positioned simply to reverse the logistics of getting their products to customers when it comes time to retrieve used and obsolete equipment." Hilton continued, "In 2003 when Dell began offering take-back services as part of sales agreements, sales to certain business customers increased three times. But how the EU's WEEE directive will play out depends on location . . . I'm not sure a federal mandate would work in the U.S. Community approaches in the U.S. can be awfully different."

Virtually without exception, e-waste legislation has been controversial wherever it has been proposed. Because collecting and processing e-waste is so much more expensive than any other kind of trash handled by municipal waste systems, the question of who will pay and how such funds will be collected is inevitably the biggest hurdle. Given the quantities of e-waste and its complexity, it's not surprising that Scott Cassell of the University of Massachusetts's Product Stewardship Institute described "e-products . . . as unfunded mandates" that "are sent out and end up on local governments' doorstep as their responsibility." "Some places," Cassell remarked, "are scared to death to collect because they're afraid of what will come in the door."[72]

California's electronics recycling regulation, established by Senate Bills 20 and 50, went into effect on January 1, 2005.[73] "This is an issue that's been talked about in my district for at least five or six years," said California State Senator Byron Sher, who was the lead sponsor of these bills. "As with many issues, the impetus came from a growing concern on the part of the public and environmental groups in our area over the improper and illegal disposal of electronic waste," Sher's chief of staff Kip Lipper told me. The big challenge, said Sher, was figuring out how to manage the cost of handling toxic materials that can't be dumped in California landfills. "The real fight is who should have more of an obligation in terms of disposal costs, the consumer or industry," said Lipper.

California's bills went through several iterations, with both manufacturers and environmental advocates weighing in with their concerns. As with all legislation, what emerged was a compromise, leaving both sides wanting more. "The bill was radically changed to make a deal with industry," says Ted Smith.

Initially, the California program is collecting computers and TVs with CRTs. In July 2005, LCD and plasma display screens—TVs, computers, laptops—and anything else with a screen measuring over four inches diagonally were added. A fee of six to ten dollars (determined by screen size) is charged for computers and televisions and the money collected will go toward supporting the recycling program. This fee—an advance recovery fee—is being added to the purchase price of new equipment (hence, the Staples notice to California customers mentioned earlier). Thus far California's program requires taking your old PC or TV to a collection point rather than leaving it on the curb, but the program is designed to build on existing trash and recycling collection infrastructure, so curbside collection of e-waste could happen (although it's not yet on the official agenda).

With the goal of diverting all toxic high-tech trash from California's waste stream, more items may be added to the list of electronics collected. Ron Baker, an information officer with California's Department of Toxic Substances Control, told me in 2004 that the state "plans to test everything" electronic—including video-game players and microwaves—to determine the equipment's heavy metal and other toxic contents. California has also passed a separate law requiring all California cell phone retailers to submit to the state a plan for recycling used phones. The legislation will also prohibit phones that don't meet the European Union's materials standards from being sold in California.

To participate in the California program, electronics recyclers must register with the state and meet certain criteria, which include ensuring that the material collected will be handled safely and in an environmentally sound way. California's law also aims to prevent export of hazardous waste. It requires exporters to register with the state and confirm that no e-waste export will take place unless the receiving country can legally accept the e-waste and will handle it as would be required in California. These requirements, which mirror the European Union's WEEE directive, are all more stringent than any federal regulation. But given the murky channels much e-waste travels, it remains to be seen how effective such rules will be.

|||||| |||| |

Some of those advocating for producer responsibility wonder how well the California program—as now structured—will work. "I view the advanced recovery fee as a problem," said Tim Rudnicki, a government affairs advocate who works on e-waste issues in Minnesota, at the 2004 E-Scrap conference. "If it's too big, it could pose a problem for retailers, and if it leaves a gap between the cost of processing waste and what the fee collects, that gap would be made up by taxpayers . . . On e-waste issues, the U.S. continues to be an international laggard," said Rudnicki, adding his voice to the chorus of consensus on that point.[74] Rudnicki believes Minnesota should tackle e-waste by prohibiting the disposal in solid waste of "virtually everything with a circuit board and a cord." The ban would, in theory, force manufacturers to rethink product design, something an advance recovery fee would not necessarily encourage.

Beginning in July 2005, CRTs were banned from Minnesota's solid waste, and Hennepin County— home to Minneapolis and St. Paul—now has a recycling program that collects consumer electronics from residents free of charge. A motivating factor behind this legislature is the desire to curb pollution of Minnesota's thousands of lakes. Rudnicki points out that almost every body of water in Minnesota is considered impaired—in other words, polluted. And the toxics contained in the high-tech trash Minnesota is expected to generate between 2004 and 2015 is considerable: 565,000 tons of leaded glass, 93 tons of cadmium, and 22 tons of mercury.[75]

Watery Maine's interest in ridding its landfills, smokestacks, and other municipal waste repositories of the toxics associated with e-waste is prompted by similar concerns. When heavy metals leach out of landfills they often enter ground- or surface water, and airborne toxics often move with rain and snow. Bioaccumulative toxins and some metals—mercury, for example—often show up first in fish and fish-eating birds. And fish are one of the ways that such toxins creep through the food web and enter our bodies. Forty-five states have now issued mercury advisories for their anglers.[76]

Bans on certain hazardous materials, judicious use of the precautionary principle, and requiring more manufacturer responsibility for disposal must all come into play to achieve any meaningful reduction of toxics in consumer products. But when compared to U.S. federal policy, California's legislation (which covers about 12 percent of the U.S. population)—along with that of other states—represents something of a great leap forward in preventing at least some used high-tech electronics from turning into truly toxic trash.

||||| ||| |

"There's very little competition between Maine and California for anything," says Jon Hinck, staff attorney for the Natural Resources Council of Maine. But where e-waste and related toxics are concerned, Maine is giving the West Coast trendsetter something to measure itself against. Maine is the first state to pass e-waste legislation incorporating extended producer responsibility principles, and it has also passed legislation, comparable to California's, banning the in-state sale of products containing certain brominated flame retardants. "We passed the legislation with an understanding that we're doing it first," says Hinck. "The state motto is, after all, 'Dirigo' which means, 'I lead.'"[77] But, Hinck tells me, "the person you really want to talk to is Maine state senator Tom Sawyer," who sponsored the legislation.

When I reach Sawyer, he is happy to talk despite the fact that I'm catching him less than a week before election day. "I spent twenty-seven years building up the state of Maine's largest solid waste recycling facility," Sawyer tells me. "I operated a sanitary landfill—Sawyer Environmental Services—and sold that in 1996 to a company in Vermont. I first ran for the Maine Senate in 2000, and in my first term served on the labor committee and the natural resources committee . . . In 2002, when I was ranking Republican senator on the natural resources committee, the electronics recycling bills first came under discussion." So, he says, "I get to wear two hats. I'm interested in environmental matters and I have nearly three decades experience in business."

As Hinck tells the story, "When we started looking around for abandoned electronic waste in Maine, it didn't take long to find it—some was stashed in cabins in the woods. *Exporting Harm* was shown in Maine and people only needed to see it once. A core issue for us was that a large volume of hazardous material is going places with low safety standards and low wages—going to the laps of poor people in the Third World. But how was a group of policy makers from a small state going to try to head this off?

"One of the other drivers of this in Maine," Senator Sawyer tells me, "is that 60 percent of municipal solid waste is incinerated for energy recovery. And we are worried about the incineration of toxics. A minority of waste is landfilled in Maine. What takes place upwind of us is very important."

In 2002 Maine passed a bill that prohibits any waste that could become hazardous, including electronics—especially computers with CRTs—from being incinerated or dumped at local waste transfer stations. The following year the state adopted a ban on CRT disposal, which provided the jumping-off point for the 2004 legislation banning electronic devices with CRTs and flat-panel display screens over four inches from state landfills and incinerators. This ban—which doesn't include cell phones—goes into effect in 2006, so Maine urgently needs to create a safe disposal route for its high-tech trash.

Maine's 2004 e-waste bill began as an advance recovery fee bill—like the one California has passed—and initially found favor with electronics manufacturers because going this route would relieve the companies of financial responsibility. "We found that 73 percent of Mainers supported the idea of an e-waste collection system but the general feeling was that they didn't want to pay for it," says Hinck. An advance recovery fee, it was felt, would seem like a second sales tax in Maine and would put retailers in a poor position to compete with those in neighboring New Hampshire, which has no sales tax. And those who think like Sawyer and Hinck felt such a fee would do little to encourage more environmentally friendly design. So Sawyer decided to promote producer responsibility legislation instead.

"All the major electronics manufacturers—with the exception of HP and Dell—showed up personally to oppose the Maine bill," says Hinck, who has the state hearings on videotape. "It was not all love and roses. It became very contentious," says Sawyer.

To oppose Maine's legislation Apple hired a lobbyist for two weeks who, according to the Natural Resources Council of Maine, was paid over $8,600. Other opposition came from the Association of Home Appliance

Manufacturers, Hitachi, IBM, JVC Americas, Matsushita of America, Mitsubishi, Panasonic, Sanyo, Sharp, and Sony—an impressive showing considering that Maine has a population of only 1.3 million. A local AFL-CIO affiliated union also initially opposed the legislation, saying it was "an anti-job bill" but later changed its mind, calling the bill "in all likelihood" beneficial to its members and the state of Maine.

"We were not so naïve as to assume that this might not result in slightly higher prices," says Sawyer, who believes any such increase will be minimal. "Maine is a small market, and I think we protected our retailers," he says. "You can't aim for perfection but we did the right thing from a recycling and environmental point of view, and did it in a free market manner."

The bill that passed will make consumers responsible for taking their used electronics to a local or reasonably close drop-off site, and many municipalities already have collection points set up. The recycling fee, if any, will be small. The municipality will ship equipment from collection points to a waste consolidator. From that point on the manufacturer will pay for the cost of recycling its own products and a proportionate share—based on market share—of orphaned waste.

So why did a bill with so many extended producer responsibility features succeed in Maine? I asked Hinck.

"It's basically business clout," he said. "Such bills stand little chance in any state with a significant number of employees of companies who would be opponents. There are zero high-tech companies who are employers in Maine," he explained. "We also have a Democratic governor, a Democratic state house and senate. They're not wild-eyed environmentalists, but their constituencies include these concerns."

And as Sawyer points out, the Maine house in 2004 was completely bipartisan, eighteen Democrats and seventeen Republicans. "All it takes is for one person to have a cold and the state of Maine changes relationships," says Sawyer. "And with all lack of humbleness, it helped to have a lifelong business Republican leading the cause."

Plus, says Sawyer jovially, "we have the saying, 'As goes Maine, so goes the nation,' a moniker that we've kept. We're getting a reputation as more environmentally conscious. Tourism is our number one industry, so we need to focus on our environment. Whatever passes in Maine, we'll probably see elsewhere."

▌▐║▌║

Once people realize old electronics are not collectibles and are given an opportunity to dispose of them safely, they're eager to get rid of them. In the spring of 2003 the citizens of Denver set a record for one-day collections by delivering over two hundred tons of used electronics to a Dell take-back event. The event created a mile-long backup onto the interstate, with a quarter-mile backup of cars waiting to get in before it opened. "We've learned from collection events how much very, very old stuff there is," said Pat Nathan, Dell's sustainable business director who described the event at the E-Scrap 2003 conference in Orlando, Florida.

But simply creating a new set of curbside bins won't do it. The products themselves will have to change in ways that extend the life of their hardware, so they contain fewer toxics, and that make them easier to recycle. These changes are happening—but slowly. In the United States that nifty new computer, iPod, DVD player, or Blackberry does not yet come with end-of-life instructions or any materials information. So, confused about what to do with that clunky old PC, most Americans are shoving them into closets, basements, and attics or are putting them out on curbs—as I've seen in my neighborhood—with hopeful paper tags saying "FREE!"

"We need something," says Ted Smith of the Silicon Valley Toxics Coalition, "that will create a burning desire or at least a strategy to design better products." Keeping old computers out of landfills will require regulation but also "a fundamental paradigm shift," says Jim Puckett of Basel Action Network.[78] Unless we make this shift in a holistic way—addressing the hazardous materials and ecologically burdensome design throughout high-tech electronics' life cycle—the e-waste will continue to pile up, wasting resources and polluting. Part of this shift includes electronics recycling and reuse itself, because unless those costs are fully addressed, the computer you thought you'd so responsibly left with a recycler may well end up as hazardous waste in some of the world's poorest communities.

CHAPTER SEVEN

Not in Our Backyard

Exporting Electronic Waste

On a sunny afternoon in March 2004, in the darkened conference room of an office in downtown Seattle, Jim Puckett, director of the Basel Action Network, shows his staff video footage he filmed a week or so before in Taizhou, in Zheijiang Province in southern China. The film is unedited, but the images are powerful. One of its sounds still rings in my ears: metal being pounded by hand. It sounds like a blacksmith's shop—an echo of brute force and simple tools that predates the Industrial Revolution. This is not a sound anyone would associate with the wired side of the digital divide.

The film shows a parade of open trucks piled high with cargo transferred from docked container ships—scrapped electronics sent overseas by richer countries for inexpensive, labor-intensive recycling. Computers, printers, office-sized copiers, old transformers, cables, and lighting units are all jumbled up with other less readily identifiable items. The trucks dump their loads on the ground in what looks like an enormous parking lot. We see pools of dark oily liquid seeping out from under mounds of the junked machinery.

We then see people in backyard workshops pounding metal, banging apart computers, sorting plastics and wires, and tossing bits of what Puckett tells us are copper and aluminum into open basinlike braziers—like big woks—and simple brick furnaces that look like chimneys. Smoke and dust billow out. In smaller outdoor workshops we see chopped-up circuit boards being roasted in uncovered pans to melt away plastics and isolate valuable metals. The scene reminds me of the open-air market restaurants I've visited in China, but instead of pan-fried noodles the fare here is seared semiconductor, replete with lead, brominated flame retardants, and plasticizers.

We also see dormitories where workers live in tiny rooms only steps from where this electronics dismantling and low-tech smelting takes place. Plastic washbasins sit outdoors on concrete blocks, and laundry hangs within sight of the work areas. The camera zooms in on a barcoded identification tag on the side of a Dell computer sitting out in a yard somewhere in Taizhou. The tag reads: "Property of the Internal Revenue Service."

"The volume was amazing," Puckett tells me over lunch. "It was arriving twenty-four hours a day, and there was so much scrap that one truck was loaded at the docks every two minutes." Much of this scrap had just arrived from Korea and Japan, and "on any given hour, hundreds of such trucks are moving down the streets of Taizhou," report Puckett and staff members of Greenpeace China who accompanied him on this trip. "We watched the trucks dump the e-waste in yards where former farmers were sitting there with blow torches and chisels, separating the stuff." This kind of used electronics disassembly, Puckett and his colleagues learned from interviews, has increased dramatically in Taizhou over the past couple of years.

The trip to Taizhou was not Jim Puckett's first encounter with e-waste in China. In 2002, Basel Action Network released a film called *Exporting Harm: The High Tech Trashing of Asia.* Accompanied by a report, and produced with help from Silicon Valley Toxics Coalition, Greenpeace China, and others, the film was responsible for the first widely disseminated graphic documentation of what happens to e-waste when it leaves the industrialized West.

I saw *Exporting Harm* in March 2002, the first time it was shown to an audience of electronics recyclers and high-tech manufacturers. Its pictures of a landscape ruined by high-tech trash and of children picking through the electronic detritus were watched in appalled silence. I've since seen the film screened for other professional audiences, and each time the room falls completely silent. When I've asked U.S. state policy makers what prompted them to take action on e-waste issues, many of them have cited *Exporting Harm*. No one wants to see their state—or company's— equipment ID tags on electronics lying in slag heaps on a riverbank or being dismantled by a woman whose child sits at her feet while toxic dust flies. The film, said Lauren Roman, who is now executive vice-president of the materials recovery company MaSeR, shortly after the film's pre- mier, "caused a paradigm shift in electronics recycling."[1]

Like the video I saw in Puckett's office, *Exporting Harm* was shot in southern China—but rather than in Taizhou, it was filmed in Guangdong Province, about half a day's drive north of Hong Kong. Much of the film was shot in and around the village of Guiyu, which had become a major repository for the e-waste that Puckett calls the "effluent of the affluent."[2] Looking at the pictures of burning wires, smoking molten plastics, and workers squatting amid voluminous piles of high-tech scrap, it took no leap of imagination to understand why Puckett describes this scene as the "underbelly of our consumptive cyberage lifestyle."[3]

Exporting Harm showed enormous uncontained sliding mounds of trashed electronics piled throughout Guiyu. The heaps of discarded com- puter parts—monitors, printers, toner cartridges, keyboards, circuit boards, cell phones, wires, and plastic cases—rose like dunes above a riverbank, their toxic ooze—that contains cadmium, copper, lead, PBDEs, and numerous persistent organic compounds—seeping into and poisoning the local water supply. Junked computer equipment lines one bank of the Lianjiang River as far as the eye can see. Some waste has simply been dumped in the river. So much trash has been left on the riverbank that some people dig old circuit boards out of the riverbed so they can recover the metal embedded within.[4]

In the mid-1990s not long after e-waste began arriving, the ground- water in Guiyu and neighboring villages became undrinkable. The entire village must now have its drinking water trucked in because local supplies have been fouled by high-tech trash. But given the expense of buying potable water, residents still wash dishes in contaminated groundwater. Children still play and swim in the toxic river water.[5] River fish supply food for the local community. Water samples taken from the Lianjiang River in 2000 showed levels of lead 2,400 times higher than levels deemed safe by the World Health Organization (WHO). Samples taken by Basel Action Network in 2001 from the same location contained lead levels 190 times higher than WHO safety standards.[6] By the late 1990s, according to a report prepared by Greenpeace China and Sun Yat-sen University's anthropology department, Guiyu had become "the largest and most concentrated site of electronic waste trade in China."[7]

Because proper disposal and recycling of obsolete electronics is difficult, labor intensive, and therefore expensive, and because communities throughout the developed world have said, "Don't dump it here," large quantities of this waste are shipped overseas to developing countries where labor is so cheap and environmental laws are often lax. For years—probably beginning in the late 1980s[8]—huge containers of old computer equipment have been shipped out of the United States, Japan, Korea, the European Union, and other developed nations and sent to less wealthy countries by recyclers who are actually brokers or dealers in e-waste. In addition to China, this waste has been going to many other countries, including India, Pakistan, and Nigeria.

▌▌▐▌▌▌▌

Exporting Harm shows workshops throughout Guiyu where workers with unprotected hands pick through tangles of circuit boards, cathode ray tubes (CRTs), ink cartridges, metal frames, cables, circuit boards, and other parts that lie, unsorted, on the ground. A man wearing no safety clothing or respiratory protection whatsoever crouches amid a pile of dismantled circuit boards, taking them apart by hand, sneezing at the dust. Several workers sit literally on top of the trashed electronics digging through the piles that rise as high as the adjacent workshed rooftops. Some piles of e-waste reach to second-story windows.

The camera zooms in on the ID tags of a pile of computers: "Property

Where it's from

of the City of Los Angeles," "State of California Medical Facility," "L.A. Unified School District—Information Technology Division." Some of the other discarded equipment Basel Action Network saw in Guiyu belonged to the Kentucky Department of Education, the Racine, Wisconsin school district, to a bank in Chicago, and to the U.S. Defense Intelligence Agency.

"People think that they did something right when they took their computer to a recycler, but it went all wrong," says Puckett. "Any of this waste could be my computer or your computer. There's a great frustration that the consumer has so little control over what happens.[9]

The camera pans through living quarters—some look like small sheds—that stand next to mounds of ash left over after wires have been burnt. People are banging at old CRTs by hand, smoking while they work, and picking up parts barehanded. There is broken monitor glass on the ground. Another picture shows an irrigation canal used as a dump for unwanted CRT glass that contains four to eight pounds of lead per monitor. A small child squats on ash-blackened ground between riverside puddles and rivulets, surrounded by a morass of electronics trash.

Elsewhere in Guiyu, workers wearing no protective clothing and using simple hand tools crack open and melt computer parts over open flames to extract reusable and precious metals. Plastic-coated copper wires are burned to retrieve their metal. The ground and houses of whole villages are coated in the resulting ash. According to Puckett, these ash piles are full of dioxins—some of the most toxic chemicals known—resulting from low-temperature burning of wires' PVC coating. A scene shot at night in a dry streambed shows what looks to be a bonfire of old electronics. Its visibly noxious black smoke wobbles above the orange flames.

In some parts of Guiyu over 80 percent of the residents of this former farming region are now involved in e-waste processing.[10] Guiyu's electronics processing workforce is estimated to be about a hundred thousand—much of it made up of migrant laborers from surrounding and more distant provinces.[11] According to the Western press and both Chinese university and NGO researchers, conditions in these workers' rural villages are so poor that even the primitive electronic scrap industry in Guiyu offers an improvement in income. But when pressed, many electronics scrap workers bemoaned the pollution and the many unsavory aspects of the opportunistic waste trade, saying they didn't want their children—many of whom were growing up surrounded by e-waste—to engage in this work.

Before e-waste processing took root in Guiyu, the area had been agricultural. A glossy 2004 brochure from the Guiyu equivalent of the chamber of commerce shows a picture of a plump green watermelon on the vine. But farming in Guiyu has been abandoned for the more lucrative scrap electronics work—and also for fear of pollution's effects. "The farmers dared not consume the food they produced, and only sold to the outsiders who did not know the pollution," report Greenpeace China and Sun Yat-sen University observers.[12]

How it affects the U.S.

In small outdoor work areas, workers sit in front of small braziers where circuit boards containing plastics and sometimes fiberglass, impregnated with chemicals and metals, are melted down. When the boards are soft and liquid, the solder and chips are plucked from the toxic soup. The lead-based solder is collected for metals dealers, and the microchips are transferred to another area of Guiyu for further processing. There, beside a river, chips are bathed in open buckets and barrels of acid to extract the gold in an ages-old process called "aqua regia" (Merriam-Webster's dictionary dates the phrase to 1612). The process uses a mixture of pure nitric and hydrochloric acids to precipitate the gold out of the plastic and other metals. Puffs of caustic steam—gas likely to contain chlorine and sulfur dioxide—rise from the liquid's surface. After the gold has been reclaimed, the waste liquids—pure acids and metals—are routinely dumped in the river. Basel Action Network tested the puddles around these acid-stripping buckets with litmus paper and found a pH level of 0—pure acid.

Computer monitor screens are smashed by hand, their copper yokes extracted and the shattered leaded glass tossed into a nearby irrigation canal or riverside dump. Plastic is chipped and separated, often by children. With bare hands and no respirators workers take apart plastic cartridges and, using small brushes, sweep out the ink and toner, clouds of chemical vapor and particles of carbon black swirling in front of their faces.* The plastic housings of the cartridges often end up dumped out in the open along with CRT glass and other abandoned plastics. By the mid-1990s over 1 million tons of e-waste were arriving in the ports of Shenzhen, Guangzhou, and Nanhai to be processed in Guiyu each year—scrap that included over 150,000 tons of plastic and over 200,000 tons of metal.[13]

Plastics must be sorted by type and color. "Since workers do not have necessary equipment for inspection, they normally have to distinguish by smell through burning the plastic," report researchers from Greenpeace China and Sun Yat-sen University. One of the workers interviewed by Greenpeace listed just some of the plastics he looks for: "ABS, PVC, PC, PS, PPO, PP, POM, MMA."[14] After burn- and sniff-testing, the plastics (impregnated with brominated flame retardants and which can create dioxins and furans when burned at low temperatures) are cleaned to remove any labels. Large pieces of plastic are sliced, crushed, and then washed and dried before being further crushed into a powder. This powder will later be melted into thin rods that are cut into granules. These are sold to factories where they are used to make generally low-quality plastic products, including plastic flowers—a regional specialty in this part of southern China.

A sample of sediment taken in 2001 from the banks of the Lianjiang River where computer parts are dumped had 1,338 times the amount of chromium deemed safe by the U.S. Environmental Protection Agency.[15] Samples taken in 2005 from Guiyu's wastewater channels found levels of copper, lead, tin, and cadmium 400 to 600 times higher than what would be considered normal and safe and antimony levels 200 times higher than is considered safe.[16] These samples also found PCBs, polycyclic aromatic hydrocarbons (PAHs), brominated flame retardants, nonyphenols, phthalates, and triphenyl phosphates (TPPs), all synthetic chemicals used in plastics. Nonyphenols are endocrine disruptors, as are PCBs, which are also probable carcinogens and along with phthalates are known to be toxic to the reproductive system. TPP is known to be toxic to aquatic life, as are

---

* Carbon black is considered a possible carcinogen by the International Agency for Research on Cancer.

PAHs, which have also been linked to certain cancers in humans.[17] "This is a city of rubbish. The air is polluted. The water is undrinkable," said one plastics separator the Greenpeace researchers interviewed.[18]

Fumes from the burning metals and plastics—which include lead, PVCs, and PAHs, all of which are persistent toxics—foul the local air and send their toxics into the global air stream. Polluted air from China, containing these chemical dusts and particles, blows east, eventually arriving on the west coast of the United States before making its way inland. Condensation and precipitation along the way brings the pollutants down to earth where they begin their journey through plants and animals.[19]

"TRACTOR-TRAILER LOADS PREFERRED"

From the mid-1990s until about 2001, some 50 to 80 percent of the electronics collected for recycling in the western half of the United States (and perhaps even more of the country) was being exported for cheap dismantling overseas, predominantly to China and Southeast Asia.[20] In 2005 BAN and the Silicon Valley Toxics Coalition found that the percentage of e-waste exported had not changed. But destinations for e-waste—from the United States, Europe, Japan, Korea, and possibly elsewhere—also include India, Pakistan, Bangladesh, Malaysia, the Philippines, Vietnam, Russia, eastern Europe (particularly Croatia), Africa (especially Nigeria), and possibly Jordan and other Middle Eastern countries. These shipments take place despite the existence of international agreements designed to curtail the export of hazardous waste.

Because the U.S. does not monitor this traffic, the scant 10 percent of used computers that the EPA estimates go to U.S. recyclers includes those that may ultimately be exported as some form of e-waste. In 2002 enough e-waste was exported to create "a pile one acre square with a height of 674 feet"—over twice as high as the Statue of Liberty and nearly 70 feet higher than the Space Needle in Seattle.[21]

Because the United States lacks an official mechanism for quantifying electronics recycled, let alone for tracking e-waste exports, some waste management and high-tech electronics industry professionals have expressed skepticism about the reported amount of high-tech trash making its way overseas. At a meeting of Oregon's Committee on Electronic Product Stewardship in the spring of 2004, I heard questions raised about the 50 to 80 percent U.S. export figure cited in *Exporting Harm*. Figures like these have a way of taking on a life of their own, especially in an arena like e-waste where all volume estimates are just that—estimates. Given the paucity of national systems to handle the waste, it's hard to know precisely what domestic recyclers—whether in the United States, Europe, Japan, Taiwan, or elsewhere—are actually receiving, how much e-waste is going to landfills and incinerators, and how much is being exported. In hopes of achieving some kind of clarity, I decided to see if I could track down the source of the 50 to 80 percent U.S. e-waste export estimate. I did this by calling Mike Magliaro, vice president of Life Cycle Partners, a Massachusetts electronics recycler, who is cited as the source for the figures used in *Exporting Harm*.

"Back in the middle 1990s, when I was working with another company, I asked my partner, 'What do they do with this stuff?'" Magliaro recalls. "About 80 percent of the electronic scrap that the company was handling between 2000 and 2002 got exported," Magliaro tells me. That was "pretty much business as usual," he says. "Export of e-waste on the West Coast is a massive issue. Huge volumes were shipped out between 1997 and 1999," agrees Jerry Powell, publisher of *E-Scrap News* who has been covering the recycling business through his *Resource Recycling* publications for over twenty years.[22] Many e-waste watchers say these exports continue unabated.

The stories I heard Mark Dallura of Chase Electronics, a recycling company based in Darby, Pennsylvania, tell in March 2002, at the EPR2 conference in Washington, D.C., dovetailed with what Magliaro told me. What Chase Electronics does, according to a description that was on the company's Web site, is "Purchase personal computers . . . in large quantities—tractor-trailer loads preferred," and "Dispose of computer scrap/electronic scrap by removal, outright purchase or re-marketing."[23]

What Chase—like many other e-scrap dealers—actually does, rather than recycle by dismantling equipment for materials recovery and reuse, is to act as a broker for used high-tech equipment and its components. As Dallura described it, Chase ships containers full of loose—rather than shrink-wrapped and pallet-bound—e-waste to Taiwan, Hong Kong, China, Vietnam, Russia, and other destinations. "We often ship out a thousand containers a month with an average load of about 47,000 pounds," said Dallura.

"There are different prices for the e-waste, depending on whether it's coming from the East Coast, West Coast or Midwest. Trucking costs have to be factored in. Loads of scrap may get sold two or three times by trading companies and investors in China," he explained. A container full of computer monitors he'd shipped recently brought a fee of $2,600, said Dallura. On his recent shipping schedule there had been four container vessels. Two had gone to Hong Kong and the other two to Nanhai, China, bearing mainframe computers not covered by China's ban on electronic waste.

"Mainframes were rich in gold and drove the market. Now the market is driven by the value of copper. But circuit boards are rich in gold," said Dallura at the 2002 conference. "How much the Chinese recyclers make depends on the state of the metals markets. But they've covered prices to make sure they make a profit.

"Older terminals and monitors that have shredded fiberglass in them often end up in landfill because it's really nasty material to work with. I won't accept keyboards and printers, it's pure crap. The only way you can get the plastic off is with prison labor," said Dallura, alluding to the time-consuming process that pushes this work into cheap labor markets both in the United States and abroad.

At Chinese customs, Dallura said, the containers are X-rayed to see if the load contains any whole, presumably new equipment that may have "fallen off a truck" somewhere. "There is some smuggling and bribery," he said. "Whole Mercedes have been smuggled in containers of waste," he said, describing how rolls of hundred-dollar bills were sometimes taped inside equipment. "I could care less where they go," Dallura later told a reporter from the *Washington Post* who asked how he felt about shipping loads of e-waste to China. "My job is to make money."[24]

"WORK AS USUAL AMONG THE WASTE"[25]

Some of this e-waste ends up in India, where as of 2003 most of the country's computer waste was estimated to be imported, rather than created by high-tech equipment used domestically.[26] The *Exporting Harm* report included photographs from India that shows women in New Delhi, draped in saris, sitting on the ground surrounded by old circuit boards, extracting the parts with pliers and wire cutters. Other pictures show a young boy taking apart a computer, a mound of previously dismantled computers rising way above his head.

A 2003 report on e-waste in India by Toxics Link, a New Delhi–based nonprofit, says that the "existence of international as well as local trade networks and mushrooming of importers of old computers in far flung areas like Darjeeling, Kerala, Kochi, etc., indicate the huge import of obsolete technology in India. It is reported that about 30 metric tons (MT) of imported e-waste landed at Ahmedabad port. It consisted of monitors, printers, keyboards, CPUs, typewriters, projectors, etc. Out of this, 20 MT was pure scrap and 10 MT was in reusable condition. That the whole amount will occupy 30 full size trucks gives an idea of the volume of waste imported. The exporting country in this case was the USA."[27]

According to the Toxics Link report, computer and other high-tech electronics disassembly and recycling in India takes place in low-tech workshops much like those in southern China. Circuit boards are burned in the open to recover copper films. Wires coated with PVC plastic are also burned in the open, as are capacitors and condensers, some of which could easily contain PCBs. Hard drives and floppy disks are melted, and gold is covered in rudimentary acid baths. A photograph in the report shows a man dismantling a circuit board by hand. He's holding the board in place with his feet—he's wearing thong sandals and no socks—and is bending over to reach the board.[28] Another photo shows a young woman sitting on her heels, surrounded by dismantled and discarded computers, working on a piece of equipment with unprotected hands.[29]

"No sophisticated machinery or protective gear is used for the extraction of different materials," reports Toxics Link, which surveyed ten electronics recycling areas around New Delhi. "All the work is done by bare hands and only with the help of hammers and screwdrivers. Most often, child labourers are employed by the unit's owners for the reclamation of lead and pulling out ICs [chips]."[30]

Unlike other e-waste destinations like Guiyu, where most of the leaded CRT glass was simply discarded, in India some CRTs from computer monitors were either refurbished for another use—for local television and video-game manufacture—or smashed and sold to a dealer who sells it to Indian glass manufacturers. But "unlike China," writes Toxics Link, "nothing is dumped in open fields, riverbanks or ponds. Though not dumped," the report continues, "the hazards involved in product recycling cause environmental damage to an unredeemable extent."[31]

Accounts of Indian e-waste recycling published in the *San Jose Mercury News* confirm what Toxics Link reports. A 2003 article describes work at one New Delhi electronics recycling operation:

[A] half-dozen workers scurry about dipping circuit boards in and out of blue plastic drums filled with acid, stripping the boards of their last remnants of copper and traces of silver. None of the workers wears a mask to ward off noxious fumes, and only one has thin yellow gloves to protect his skin from the toxic brew. When the acid is depleted, the men dump it into the open sewer lining a rutted dirt side-road in the Mandoli industrial area, a collection of small, decaying factories in the northeastern corner of India's capital.

Maheshwari [the proprietor of the recycling shop] says he ignores city regulations and burns the bare plastic boards in the open air, just like the 10 rival scrap yards doing the same work in the area. He boasts of procuring his scrap from North America, South Africa and Hong Kong, which he processes along with computer waste generated throughout India.[32] *Where Waste comes from*

Open burning of e-waste that left high levels of cadmium, copper, lead, zinc, and heavy metals dust resulting from CRT disassembly and storage were still in evidence when researchers from Greenpeace International visited e-waste processing areas of New Delhi in 2005.[33]

TALKING E-WASTE IN CHINA

"We started hearing about these exports about five or six years ago," Jim Puckett tells me in March 2004. "But where do you go look? China's a huge country," he quips. So Puckett enlisted the help of the young, bright, and energetic staff of Greenpeace China, which has offices in Beijing and also in Hong Kong, about a days' drive from where much of the imported e-waste makes landfall in China. It's staff from Greenpeace China, along with international colleagues and university researchers from Shantao and Hong Kong, who have compiled most of what we know about the fate of this high-tech trash.

Some of the first internationally available reports of the e-waste trade appeared in business journals of the metal scrap industry. The metal scrap business has boomed astonishingly in China since the late 1990s. As of 2004 China was the world's largest consumer and importer of scrap metals, and the United States the world's largest producer of such scrap.[34] Statistically, electronic scrap that contains metal—as virtually all of it does—is bundled in with other metal scrap.

At the same time, scrap and waste of all kinds have become one of the United States' largest exports. "In November, the United States had a record trade deficit of $5.8 billion in advanced technology products. For the most recent 12 months, the deficit was $36.9 billion, also a record. And where is the strength? The trade surplus in what the government calls 'scrap and waste' is rising. The 12-month total of $8.4 billion in such exports is up 31 percent from a year earlier," wrote Floyd Norris in a January 2005 New York Times article titled "U.S. Tech Exports Slide but Trash Sales Are Up." Compared to 1999, wrote Norris, "Exports of advanced technology products are down 21 percent, while those of scrap and waste are up 135 percent."[35]

Most of the electronics processed in places like Guiyu and Taizhou comes from Japan and the United States by way of Hong Kong and Taiwan.[36] After arriving in Guiyu the loads of scrap electronics are purchased by individual processors. The big loads are typically divided up by type of equipment, component, or material (metals, plastics, circuit boards, wires) for specific materials recovery. This means that sometimes a second purchase is made by processors who specialize in a certain type of material. This trade is also dispersed between large-scale scrap dealers and smaller ones, with some families engaged in Guiyu's e-waste trade operating on what could be described as a freelance basis.

A large part of what has made China's scrap processing so profitable is its abundant workforce and cheap labor. "Any job that requires 'too

much labor' is good for us," says Joe Chen, president of a metals processing plant in Nanhai, who spoke to journalist Adam Minter in 2003. "In
America it's the other way around—the last thing a scrap dealer wants is
'labor intensive.' Here 'labor intensive' is how we make money."[37] China's
larger e-waste processors employ hundreds of people—with large numbers working throughout several shifts a day. Most U.S. electronics recy- *American*
clers have fewer than twenty employees; only a few have over one hun- *recyclers*
dred.[38] The electronics recyclers I visited in the United States had
astonishingly few people on the work floor at any given time.

How much does labor cost in southern China's electronics recycling
workshops? The survey conducted by Greenpeace China and Sun Yat-sen
University found that salaries ranged from 17 to 32 yuan per day (in 2005
exchange rates, about $2 to $4 dollars a day), with women typically paid
about 5 yuan a day less than men.[39] Base salaries reported by *Scrap* magazine in 2003 for work in larger electronics recycling factories were about
$75 a month, which sometimes included room in a factory dormitory.[40]

"Until now, nobody, not even many of the reputable recyclers," say the
authors of *Exporting Harm*, "seemed to know the fate of these 'Made-in-
the-USA' wastes in Asia and what 'recycling' there really looks like. And
it was clear that many did not want to know."[41]

In April 2004 I got to hear directly from Chinese researchers who have been
studying the environmental and health impacts of electronics recycling in
southern China. I traveled to Beijing to attend the International Conference
on Electronic Waste and Extended Producer Responsibility in China, hosted
by Greenpeace China and the Chinese Society for Environmental Sciences.
That the conference was even being held and that Greenpeace was the lead
organizer seemed remarkable. That there is a Greenpeace China seemed in
itself remarkable. Nearly all of the 150 or more people at the conference—
news media, government officials, and students and representatives from
international high-tech businesses, universities, and NGOs—were Chinese.
There were a handful of Greenpeace International staff—from India, the
Netherlands, the Philippines, England—as well as attendees and presenters

from Japan, Singapore, and Switzerland. I was one of four Americans there, and the only non-Chinese journalist. It was a living example of one of those world maps produced for the Asia-Pacific region, where China sprawls across the middle of the page and Europe and the United States are squashed out on the margins.

The conference was held in a semiofficial, recently built hotel called the Foreign Experts Building, with elevators that had carpets, changed daily, with the day of the week woven in—helpful for the jet-lagged foreign expert. The hotel is on one of Beijing's new outer-ring roads and is surrounded by brand new, high-rise apartment buildings, more of which were under construction. This visit to Beijing in 2004, ten years since a previous visit, provided a powerful illustration of China's ferocious growth spurt and the country's ravenous appetite for raw materials, scrap, and new technology. Building and road construction were going on seven days a week, with work beginning at what seemed like predawn. The old hutongs of central Beijing had almost entirely disappeared. Whole neighborhoods of narrow winding streets with one-story gray brick houses with sloping tile roofs had been razed. In their place loom gargantuan shiny new buildings, large enough to house both offices and shopping complexes, and sometimes apartments, under one roof. "It looks like Las Vegas," observed a Chinese American acquaintance on her first visit to Beijing in twenty years.

Many of the new Beijing office towers are crowned with the insignia of international corporations. The roads were clogged, not by bicycles as they had been on my last visit, but by cars—nearly all new full-size vehicles. And everyone—from young women wearing jeans and trendy blouses, and whose tiny phones dangled from decorative neck cords, to elderly men wearing old cobalt blue Mao jackets—was on a cell phone.

On the plane to Beijing I perused the current issue of the *China Business Review*, which reminded me how much high-tech manufacturing is taking place in China. "Xi'an, Shaanxi and Chengdu, Sichuan have recently won major investments from foreign semiconductor manufacturers," began one article. "Intel chose Chengdu High-Tech Zone for its western China facility . . . a semiconductor packaging and assembly plant . . . reportedly the largest foreign investment project in western China," said yet another. Leading foreign investors in the industrial zone, I learned, included "Honeywell, NEC Corp., Fujitsu, Brother Industries, Royal Phillips Electronics, Motorola, Mitsubishi," and "Infineon Technology, which will set up an integrated circuit design center" there. The list of such investments went on and on. Clearly, the pictures of primitive e-waste recycling coming out of Guangzhou and Zheijiang would not sit well with Chinese business leaders engaged in attracting multimillion-dollar high-tech industry investments from international corporations. And clearly, it would serve China well to denounce this environmentally unsound treatment of e-waste.

"E-waste is a very important social problem," said Professor Wei Fusheng, commissioner of the Environment and Resource Committee with the National People's Congress, at the opening session of the conference. He condemned "seventeenth-century ways of dismantling e-waste" and said that we need to have "scientific management of e-waste to prevent pollution and protect people's lives." And, he pointed out, "if usage becomes circular, we can change waste to treasure and precious resources." But for now—the primitive e-waste processing continues.

Huo Xia of Shantao University Medical College reported on her medical investigation of the effects of persistent toxics released during e-waste processing in Guiyu since the mid-1990s. "We generally encountered a lot of difficulty from local government and workshop bosses," she said. "But we were able to determine—by working with a private hospital—that many people were suffering from respiratory and skin diseases, as well as headaches, dizziness, and various chronic gastric complaints." She showed slides of workers' hands with badly discolored fingernails and vivid rashes. Many of these symptoms, she believes, result from the heating and burning of plastics that takes place when circuit boards are softened to extract capacitors, chips, and metals—and possibly also from exposure to tainted water. Other health problems she encountered appear to result from a combination of water quality degraded by industrial contamination and by Guiyu residents' subsequent intake of insufficient quantities of expensive bottled water, which residents sometimes mix with the polluted local groundwater.

A woman interviewed by Greenpeace China and Sun Yat-sen University researchers whose work consists of "burning circuit boards" said that she knows that "burning those boards is very hazardous. The toxic fumes made me vomit when I began working on that," she said. "It made me feel dizzy. But now I get used to it already. Our boss is not bad as he gives each of us a fan. Otherwise it is very unbearable especially in summer. Our boss has also installed several ventilators in the factory."[42]

The Pearl River delta region, said Professor Ming H. Wong, chairman of the biology department at Hong Kong Baptist University and director of the university's Institute of Natural Resources and Environmental Management, has experienced the highest increase in persistent organic pollutants in China—if not the world. Conference-goers listened as Wong explained that the sediment in the river and its fish are a storehouse of these pollutants, the most potent of which are dioxins and furans. They are odorless and colorless, and their primary source is the incineration of waste containing PVCs. Once in the air these compounds are transported and dispersed through repeated cycles of condensation and evaporation. When they come down to earth they can be taken up by plants and can work their way into soil and water sources, from where they enter other plants and animals—including humans. Airborne, they can also be inhaled or be absorbed by porous plant and soil surfaces. Because these compounds resist chemical, biological and photolytic degradation, they tend to bioaccumulate, particularly in fatty tissue and lipids, and biomagnify as they work their way through the food web. Over 90 percent of humans' uptake of these chemicals comes by consuming meat, fish, dairy, and eggs, Wong says, and they can also be transmitted through breast milk. These compounds unfortunately also have a long half-life: it takes five to ten years before they disintegrate in the human body.

The open burning of e-waste is a common scene in Guiyu, said Wong. "This releases respirable particles of toxics," he told me. These are particles of heavy metals combined with chlorinated gases and dioxins, along with polybrominated diphenyl ethers and brominated dioxins. Such particles can stay permanently in many cells, he explained. "There seems to be a critical problem here with PBDEs," he added. Soil sampling that he has been doing as part of the e-waste impact study in Guiyu has found levels of polycyclic aromatic hydrocarbons and PBDEs that are ten to sixty times higher than those reported from other areas of the world with PBDE contamination.

"THE ECONOMIC PATH OF LEAST RESISTANCE"

Since 2002, when the world first got an eyeful of what happens to their old high-tech equipment in the backyards of Guiyu, the Chinese government has been trying to control the import of e-waste. The government has issued regulations requiring all imported scrap to be registered. According to these regulations, Sarah Westervelt of the Basel Action Network explained to me, "All importers of scrap must be registered with the Chinese government." The aim is to be able to track down the brokers so that the Chinese government can reject a load of toxic scrap. In the past, these brokers would suddenly disappear. But thus far these rules—and their implementation—seem to be riddled with loopholes that allow e-waste to come into the country along with other loads of scrap metal. With fees paid up front to recyclers, who also receive money at the back end from those who receive the scrap for processing, "there's so much money to be made," Jim Puckett adds.

So although as of August 2002 the Chinese government has officially banned imports of toxic e-waste, they continue to come. The crackdown has curtailed—but not eliminated—shipments from North America; what Puckett and his colleagues from Greenpeace China found in mid-February 2004 at the port of Taizhou were huge ships full of uncontainerized electronic scrap arriving from Japan and South Korea. Japanese law also forbids such exports, but some e-waste analysts estimate that over a third of Japan's e-waste is currently disappearing without being accounted for.[43]

Beginning in about 2002, following passage of its electronics recycling legislation—but also partly in response to *Exporting Harm*—the European Union began a concerted effort to monitor and crack down on export of e-waste. Yet despite this increased attention in Europe and awareness in the United States, and China's efforts to control the influx of e-waste exports, the ships full of high-tech trash continue to sail. And there is concern that brokers continue to find new ports of call for this cyberage waste.

In an attempt to prevent e-waste from following what Puckett calls the global "economic path of least resistance,"[44] the fifteen countries of the European Union—and about fifty other countries—have signed an amendment to the Basel Convention known as the Basel Ban. The Basel Convention is an agreement that was negotiated by the countries of the United Nations Environment Programme in 1992 to regulate the international trade of hazardous waste. Countries that have ratified the agreement—of which there are now over 160—must ensure that they get written, informed consent from recipient Basel countries prior to exporting wastes, and that any imported hazardous waste is handled in an environmentally sound manner, following guidelines developed as part of the treaty. As Sarah Westervelt explained to me, the Basel Convention also prevents its parties from trading in hazardous waste with countries not party to the convention, unless allowed by a separate treaty. The only such agreement the United States has entered into is with the OECD (Organization for Economic Cooperation and Development—a coalition mostly for First World countries that works on trade as well as environmental policy issues), which allows a less restrictive trade in hazardous waste between OECD countries, regardless of whether or not they are Basel parties.

The Basel Ban, added to the Basel Convention in 1995 but not yet in full legal force, prohibits the export of hazardous waste from countries that have ratified the Basel Convention and that are also members of the European Union or OECD to any non-OECD/EU countries. The aim of the Basel Ban mirrors that of the convention: to prevent the export of hazardous waste—including that in high-tech electronics—from wealthy countries to less well-off ones. As of 2005 about fifty-five countries—including all of the EU countries—had signed and ratified the Basel Ban. As of early 2005 the United States had signed but not yet ratified the Basel Convention—the only industrialized nation not to have done so—and was not a signatory of the Basel Ban.[45]

"Europe is way ahead of us in pollution-prevention technologies and used materials collection," observes Puckett. The vision of the Basel

Convention, he explained, was to "loop responsibility for a used product back to both consumer and manufacturer. If you enjoy a product," either by using it or profiting from its manufacture and sale, "you should take responsibility for making sure it's disposed of properly."

As a member of the OECD, the United States is required to make written agreements of informed consent with a receiving non-OECD country before certain kinds of waste can be exported. But, explains Bob Tonetti of the EPA's Office of Solid Waste, "Because obsolete electronics are often capable of reuse, existing EPA policy is that they are not considered waste until a decision is made to recycle or dispose of them. Non-OECD countries cannot legally accept hazardous waste without a bilateral agreement and the U.S. currently has no such agreements. However, dismantling can occur in other OECD countries." If this seems confusing, that's because it is. These policies also make it hard to know what ultimately happens to the thousands of computers sent to developing countries like Kyrgyzstan, Georgia, Nigeria, and Ghana, ostensibly as charitable donations.

As of the fall of 2005 over thirty electronics recycling companies had signed a pledge—initiated by the Basel Action Network, Silicon Valley Toxics Coalition, and the other NGOs that make up the Computer TakeBack Campaign—agreeing not to export hazardous e-waste to poorer countries, not to use prison labor for electronics recycling, and not to landfill or incinerate such e-waste. But it was hard, when *Exporting Harm* was being researched in 2000 and 2001, says Puckett, to find "anybody that was not exporting to some degree." Export of e-waste is extremely hard to track, but "it's very easy to deny," Puckett tells me.

"Nations have agreed not to trade in slaves, endangered species, and nuclear weapons," says Westervelt. "Most of the world has agreed not to allow free trade in toxic wastes, either. But the U.S. is the one developed nation that . . . refuses to control and monitor its exports of toxic e-wastes."[46]

Unlike the EPA, Britain's Environment Agency has looked into the amount of electronic waste the United Kingdom is exporting. In a report released in April 2004, the agency found that a "wide range of electronic and electrical equipment is exported from the UK to non-OECD countries." An estimated 160,000 metric tons (352,000 pounds)—or "tonnes" in British locution—of e-waste were exported in 2003, approximately 10 to 15 percent of Britain's total electronic waste for that year. Of this waste 133,000 metric tons were high-tech and telecommunications equipment, mostly from businesses. "A further 23,000 tonnes were undeclared or grey market exports going to non-OECD destinations, among them the Indian sub-continent, West Africa, China and other countries in the Asia Pacific region. Of this amount, at least 10,000 tonnes are estimated to be used PC monitors," says the report. About a third of the United Kingdom's exported electronic waste, if not more, the report estimates, has the potential to be classified as hazardous waste (and therefore illegal to send to non-OECD/EU countries). The total monetary value of Britain's e-waste exports is estimated to be in the tens of millions of pounds.[47]

After computers and other high-tech equipment arrive at many recyclers, the most valuable items are separated out for sale, either directly to consumers of some kind, or to an electronics broker. What's left is then typically sorted by type of equipment and/or component and sold to brokers who then ship the goods to Asia, often China (or other developing countries), where the equipment is bought by electronics scrap dealers. This kind of export can also take place without the initial separation by grade or kind of equipment, in which case whole, unsorted batches of used high-tech equipment are sold to—and bought by—brokers for export "as is."

At the May 2002 International Symposium on Electronics and the Environment, Jade Lee of System Services International, an electronics recycler with operations in Taiwan, pointed out that "markets for low-grade recycled materials in the U.S." are lacking.[48] "Over 90 percent of recyclers are forced to export low-grade material to overseas markets, directly or indirectly. The majority of recyclers don't have their own facilities in Asia, therefore they deal with brokers, hence the problem with lack of direct control," said Lee.

Two years later, at the Beijing conference, Jean-Claude Vanderstraeten of Hewlett-Packard's Asia-Pacific division commented on related difficulties in regulating international traffic of e-waste. "If an electronic product was immediately classified as hazardous waste, it would be a roadblock to our take-back system because it would require transporters to be licensed to handle hazardous waste. We need take-back and recycling systems tailored to specific countries," he said. "We need the flexibility on the Basel export ban to allow recycling to be done at the highest possible environmental standards," he explained.

▌▐▌▐▌ ▌

The route of e-waste from its last domestic user to the export broker's bin is difficult to trace and the contents of those bins may be rationalized differently. However, Britain's Environment Agency, the European Union's Network for Implementation and Enforcement of International Law, Basel Action Network, Greenpeace International, and other NGOs have had some success in tracing the global itinerary of some e-waste shipments. The European Union has given the world a good beginning in regulating the hazardous materials in high-tech electronics and in setting up systems for recycling used equipment. Yet ironically, one of the EU directives' fundamental features is also facilitating the export of e-waste. Because there are no "transfrontier shipment issues within the EU," says the UK Environment Agency's waste electronics report, not much documentation is required for cargo traveling by truck within the European Union, destined for ports like Rotterdam. "The sheer volume of container traffic through the port," says the agency, "makes it difficult to detect mis-declarations" of the kind that allow e-waste to be shipped abroad—often illegally—for environmentally unsound recycling.[49]

"We export to every major country in the world. We regularly ship to Pakistan, India, Sri Lanka, Egypt, Iran, Bangladesh, Azerbaijan, Vietnam, Jordan, Iraq, Saudi Arabia, Kuwait, Syria, Germany, Spain, France, Bulgaria, Italy, Belgium, Russia, Portugal, Romania, Ukraine, Poland, Canada, Hong Kong, Bolivia, Venezuela, Columbia, Peru, Chile, Nigeria, and Ghana. We currently export over 10,000 computers and monitors each month," said an American electronics broker quoted in the UK Environment Agency report.[50] A quick Google search for used computers will find many such business descriptions. Some of these exports are for legitimate reuse, but which are not is almost impossible to tell without visiting the dealers and tracking individual shipments.

A European group called the Seaport Project spent a couple of years tracking illegal and improperly documented exports of hazardous and other waste—including e-waste—from European ports to developing nations.[51] Their final report, released in 2004, found illegal waste traffic at a number of European ports. It also discovered that "port-hopping" helped make it possible for illegal waste to travel undetected. Among the items the group found being exported were televisions leaving the Brussels area for West Africa; high-tech office equipment leaving Germany; a large load of single-use cameras (without batteries) leaving Hamburg bound for Hong Kong and China; and used and damaged computer equipment—including monitors—all marked "plastics," sailing out of Wales for Pakistan, along with loads of other scrap that included wrecked vehicles and industrial-strength electrical and other cables.[52]

E-waste is a particularly pesky category of waste to assess because, as the Seaport Project report notes, "At this stage, no profile for electric/electronic-scrap could be made, because these items are classified under a lot of different commodity/custom-codes."[53] And, cautioned the Seaport Project in its June 2004 newsletter, "The power of the network" of waste "stretches even further than the EU borders, because the Western economy is responsible for an environmentally friendly disposal of waste due to its prosperity."[54]

"Is it a Chinese issue or a U.S. issue that the final resting place for high-tech trash is a roadside ditch in the heart of China?" asks Puckett rhetorically. "It's the responsibility of electronics recyclers to find out who they're doing business with and to do the downstream audit to make sure their material is being dealt with responsibly," said Lauren Roman, at the 2002 EP2R conference on electronics and the environment.

But not all high-tech electronics dismantling that takes place in non-OECD countries happens in bare-bones backyard workshops. For example, there was Singapore-based Citiraya Industries,† which handled high-tech trash in what the *Wall Street Journal* described as "a high-tech fashion."[55] Equipped with state-of-the-art machinery and safety gear, Citiraya documented all the material it received throughout the recycling process, but part of the plant's attraction was its low labor costs. Hewlett-Packard executives quoted in the *Journal* article said that because the company was able to hire so many people to do the initial manual disassembly, Citiraya was able to recycle about 90 percent of the equipment it receives. In contrast, the rate of recycling efficiency for the automated systems in use in Europe was put at 65 percent. None of the electronics recyclers I visited or have spoken to have yet developed a purely mechanized disassembly system. Until design of high-tech products changes significantly, the first step in recycling will remain taking the stuff apart by hand.

"Every country has their own definition of hazardous and solid waste, so it can be a real morass" regulating where electronics can or cannot be sent for recycling, said the EPA's Bob Tonetti in 2002. For example, Australia has ruled that circuit boards, because they contain lead, are hazardous waste and, under Basel Convention, can't be sent to Singapore, China, Indonesia, Thailand, or India, although they can be sent from Australia to OECD countries. Another way around this problem is to rid the equipment of toxics. Things are moving in that direction, but slowly. The lead may be coming out of solder used in circuit boards, but many other toxics—for which no substitutes have yet been found—remain, and the e-waste being processed will, for quite some time, be that of older generations of equipment.

---

† In the fall of 2004, the *Wall Street Journal* reported that Citiraya had signed contracts with Intel, Nokia, and Hewlett-Packard, among other high-tech manufacturers, and had thus had gained "70 percent of the corporate market for scrapped electronics." But Citaraya has since come under investigation for corrupt financial practices and its business has taken a nosedive.

"WE SHOULD HAVE A SYSTEM FOR
ELECTRONICS LIKE WE HAVE FOR SOY SAUCE"

"E-waste exports do not lead to sustainable development," said Lai Yun of Greenpeace China at the April 2004 Beijing conference. "The long-term costs will be greater for China than the short-term benefits. It will prevent manufacturers from moving to green design. It is also a disproportionate burden on poorer communities," he says of e-waste treatment. Since the primitive recycling of electronics in China was first publicized, the government has made some improvements in Guiyu, said Lai Yun. But, he added, there is still no regulation of small backyard, e-waste dismantling workshops.

Several other Chinese-government speakers at the conference stressed the need for an economy that incorporates principles of recycling. But where reality and desired policy intersect in China seems to depend on who you ask and where you look.

"Developed countries are dumping e-waste produced in developing countries on those developing countries while developing countries are producing their own e-waste," said Howard Liu, executive director of Greenpeace China, speaking at the same conference. This put a less encouraging spin on the notion of a "circular economy" that the Chinese are touting as part of the path to sustainability.

But the Chinese know they have serious environmental problems on their hands. As Xu Shufang of the science and technology department of SEPA (China's version of the EPA) put it at the conference, "the northeastern part of China's resources are used up." And the situation is perhaps worse in rural areas, particularly in the central and southern parts of the country, said Professor Wang Weiping of Beijing's Municipal Administration Committee. "Our rivers are full of rubbish," he told conference participants. Even a short afternoon's drive outside of Beijing reinforced that observation. About a half hour beyond the central city limits, streambeds were dry or reduced to a trickle and lined with trash. The nearby dusty fields sprouted bumper crops of blowing plastic bags and other household garbage.

"We need new industry to promote the economy. A new industry that would reuse waste would be very beneficial. This would also create new job opportunities," said Xu Shufang. Chinese industries are consuming natural resources (in both raw and scrap form) at a galloping pace, but China is also now consuming ever-increasing amounts of manufactured products.

"As the world's most populous country, China also has the potential to be a top producer of e-waste as both one of the biggest producers of electronic devices and a big consumer of electronic devices," said Xu Shufang. "We should have a system for electronics like we have for soy sauce. You need a deposit for the soy sauce or vinegar bottle. This credit system promotes the idea of recycling and reuse . . . We need to encourage companies to use taxation as leverage to promote the recycling industry," he continued, explaining how taxes would be more effective than administrative rules without any financial component. I thought about how unpopular an approach described this way would be in the United States—industry opponents of e-waste legislation have been lobbying against such rules precisely by decrying recovery fees as a tax.

"China is now a middle-income country where environmental challenges are so important that everyone must take an active role," said Sven Ernedal, first secretary of the European Union's delegation to China. The European Union's WEEE and RoHS directives on electronics waste and hazardous substances and the REACH legislation on chemicals used in consumer products would have an impact on EU trade with China, Ernedal told the conference audience.

The Japanese electronics manufacturer Sony, which makes products for sale worldwide that will conform to stringent Japanese standards and to those imposed by RoHS, wants all of its suppliers—and it has four thousand in China alone—to meet these requirements by qualifying for its Green Partner program. Chinese manufacturers are scrambling to meet these standards—and those that are likely to be enacted within China as well—while the government is discussing a possible research and development fund that would help domestic companies with this work.

High-tech electronics is China's fastest growing industry, said Zhou Zhongfan of China's Environmental Science Academy. "We need to

prevent waste and pollution through design and choice of materials," she
said at the Beijing conference. But, said Zhou, "The Chinese electronics
industry is facing a lot of pressure for speed, and that is creating a great generation of waste . . . Clearly, we need to extend the life of products and to
have legislation requiring clean production." And she added, "We always
want to manage the world through ethics, but we also need regulation."

Meanwhile, China is trying to control the problems created by
imported e-waste by barring certain items from entering the country. "We
all know that smuggling is a very serious problem in China," said Liu
Fuzhong of the Chinese Electrical Appliances Association, speaking at the
conference. China must also grapple with the challenges posed by the fact
that "the most advanced cities of China and the most dilapidated areas are
so vastly different in economic situations," Huang Jian Zhong of the
Chinese Ministry of Information Industry told us.

"We need a sustainable system for e-waste," said Professor Wang
Weiping. "China has most of the technologies for dismantlement," he
said, referring to the system of manual disassembly that is in use, with few
exceptions, worldwide. But China, he pointed out, lacks the advanced
technology for the next steps in electronics recycling, technology which
is very expensive: those big mechanized machines that extract metals from
circuit boards and dismantle CRTs without the use of open fires, hammers, tweezers, and vats of acid.

"THOUSANDS OF WOMEN IN CHINA
ARE COOKING CIRCUIT BOARDS"

The export of hazardous e-waste to developing countries has taken literal
muckraking—lead by the Basel Action Network and Greenpeace China—to
expose and bring to the attention of the international public. The investiga-
tive spotlight has focused on China—and to some extent India—and the
European Union has begun a concerted effort to stop shipments of e-waste
out of European ports. But the shipments continue, slipping out of new ports
for as yet unscrutinized destinations and likely moving overland to points east
of the EU borders. Having seen what the slag heaps of e-waste looks like in
southern China, one shudders to think of such sites replicating elsewhere.

"We realize the problem is spreading and there is no sign that export is stopping," Ted Smith of the Silicon Valley Toxics Coalition told me.

Really, the only way to solve this problem, says Jim Puckett, is with legislation that would make it impossible to send e-waste to "cheap and dirty outlets" that "allow for cheap externalization." The European Union's WEEE directive does specify that e-waste can't be exported to OECD countries unless the originating country can certify that it will be dealt with under conditions equivalent to those that would be allowed at home—and under OECD rules this also means the receiving and shipping countries must have a written agreement to allow the transfer. But countries that have ratified the Basel Ban, which means all EU countries, can't ship hazardous waste to non-OECD/EU countries. China has ratified the Basel Ban amendment to the Basel Convention, but "so far it's not been effective in terms of e-waste," said Puckett. "All real solutions to e-waste lie upstream," he says. As of 2005 the United States had ratified no agreements that prohibit international trade in hazardous waste. "The United States, the most wasteful country in the world, has no export controls on e-waste whatsoever," says Puckett.

"I look at what the EPA is doing, and I think it's not much," says Mike Magliaro of Life Cycle Partners. "If the EPA said tomorrow there would be no more export of e-waste, things would change, but I don't think they want to," he said. "I disagree," Magliaro told me, "with the EPA stance about the U.S. lack of infrastructure to handle this problem. This business has more potential than people realize." Most of the American electronics recyclers surveyed by the International Association of Electronics Recyclers for their 2003 industry survey said they were operating at 50 to 75 percent capacity. Some were operating at less than half their capacity.

As Zhou Zhongfan said, better design and better choice of materials are prerequisites for making the recycling of high-tech electronics a less arduous and less toxic business. The less difficult it is, the less labor intensive it will become, and if accompanied by a proliferation of technical wherewithal to replace open fires, chisels, tongs, and buckets, the price of processing e-waste should come down. Moreover, if electronics recycling becomes less expensive, there will—one hopes—be less incentive to send this work to countries where wages are low and environmental standards can be brushed aside in the name of economic progress. But we're not there yet, especially since some of the world's largest producers of e-waste have yet to staunch the flow of these exports.

"Nothing has changed in two years," said Puckett in Beijing in 2004. "I wish we had more progress to report two years later [after *Exporting Harm*]." As we sit in this conference room," he said at the time, "thousands of women in China are cooking circuit boards from all over the world." They still are.

CHAPTER EIGHT

The Politics of Recycling

If I were to dump my old laptop in the trash that I set out on the curb in front of my house in Portland, Oregon—something that is perfectly legal—it would be carted off with the rest of the neighborhood garbage. After arriving at the solid waste site in northwest Portland, my computer might be spotted and picked out of the heap by the folks who monitor the debris for salvageable appliances. If not then taken off for reuse or recycling,* it eventually would be trucked out to a landfill in eastern Oregon regardless of its toxic contents. The same would likely hold true for a TV, or any other large piece of equipment. But any single small portable high-tech electronic device—a cell phone, portable CD player, PDA-like item, or broken iPod—and peripheral items like power cords, batteries, and ink and toner cartridges, would probably get buried and pass unnoticed through the waste stream.

---

* With the term "recycle," I'm including both dismantling a computer to recover its working parts or materials so they can be put into new products and putting an intact, working computer into the hands of someone else who can use it. Sending a computer off for recycling doesn't necessarily mean that a working computer will be dismantled for materials recovery. Most recyclers sort through incoming pieces of equipment and have a triage system to separate those that can be refurbished for reuse from those that can't.

If I took an old computer out of my closet today and shipped it back to the manufacturer according to instructions, brought it to a local used electronics collection event, or somehow got it to a reputable recycler if it was not deemed reusable or salvageable for parts, it would most likely end up in a shredder where its metals and plastics would begin their next life. The same fate would befall a computer or other circuit board–bearing piece of high-tech equipment wherever it was recycled—providing of course that it was not exported for cheap, usually unsafe recycling or dumping. What would vary is where and how the disassembly and materials recovery would occur.

Step one in preventing electronics from becoming toxic high-tech trash is knowing that a used computer or other piece of equipment should be recycled. Step two is figuring out how to do it. In Europe and Japan, where the systems and infrastructure for used electronics recycling are considerably more developed than they are in the United States, getting the stuff out of the house or office still requires effort, but there's now little mystery in what one should do with it.

For example, in the fall of 2003, Japan decided to use post offices as collection points for used electronics. This gives individual consumers about 20,000 collection sites nationwide in addition to the 250 or so centers that collect used electronics from corporate consumers. To facilitate mandated recycling, Apple computers sold in Japan after October 1, 2003, come with a "PC Recycle Label" that covers the cost of returning the computer to Apple. According to Apple, the Japanese postal service will even pick the box up from you. If all goes smoothly with implementation of the European Union's WEEE directive—and this is a question, as Germany and the United Kingdom have already asked for a delay—Europeans will bring used high-tech equipment to conveniently located municipal collection points or to an electronics retailer. In a number EU of countries, WEEE will build on existing systems for collecting and recycling used packing materials. And presumably, rollout of the directive will include additional consumer education to help ensure that all this effort isn't for naught.

Meanwhile, purchasers of large quantities of IT equipment, wherever they're located—those who buy on a corporate or institutional rather than small-business scale—are beginning to include take-back provisions in their purchase agreements, which makes recycling easy. The equipment goes back to the supplying manufacturer who has agreed to make sure the used machines are responsibly disposed of—whether it is for reuse or materials recovery. If done properly—and this is a *big* "if"—having IT equipment reused and recycled protects businesses from hazardous waste liabilities.

Based on what's going on elsewhere in the industrialized world, it seems eminently possible for manufacturers to participate in electronics recycling systems and remain profitable. But the United States continues to struggle with the concept, so for Americans consumer confusion persists. If you're lucky enough to live in a Massachusetts community that collects used computers at curbside, or in a community that has an ongoing drop-off program, once you find the relevant information your dilemma is essentially solved, though you may have to pay a service fee. The major computer manufacturers now all have recycling instructions on their Web sites but because they are designed to sell merchandise rather than to recover used goods, recycling instructions are often tricky to find. These sites can be time consuming to use and not all items are covered by these directions. I've looked without success for any specific manufacturer instructions on how to recycle a PDA (like a Palm Pilot) or a digital music player like the iPod.

This sleek little white digital music device, which appears to be the epitome of Information Age hip and is exceedingly popular—over ten million iPods have been sold in a little over three years—has become an object of controversy in the ecological sustainability department. It currently has an internal lithium battery that cannot be removed or replaced by iPod users. The only way to extend the life of the iPod beyond the tenure of its original battery is to send it back to Apple for a replacement iPod or newly installed battery. (As of late 2005, cost for a new battery was $59 plus $6.95 shipping.) One could argue that having the manufacturer replace the defunct toxic battery helps ensure that it's properly handled, and that the high price reflects the true cost of the equipment. Or one could argue, as does the Computer TakeBack Campaign—which targeted the iPod by staging protests at Apple headquarters and at the 2005 MacWorld Expo—

planned obsolescence

that a nonremovable battery that may last only eighteen months is symptomatic of a product that's environmentally irresponsible, and that Apple's battery replacement process will encourage users to toss an old iPod and replace it with a whole new one.[1]

Apple spokesperson Fletcher Cook told me in April 2005 that he could not disclose what the company does with the old iPods it receives because this is proprietary information. Cook said batteries from iBooks were recycled, but where they were sent he also was not able to disclose. All the information he could share, Cook said, was on the company Web site, which was the answer I received from another Apple press contact as well. A call to Metech International, however, whose number *is* listed on Apple's Web site for recycling information, said that they receive equipment—PCs, laptops, and iPods—both from individual Apple customers and from Apple for recycling and that they remove batteries from both laptops and iPods and send them on to an unnamed third-party recycler.

Nonremovable batteries notwithstanding, batteries are recyclable, and some U.S. retailers—Radio Shack is one—have collection bins for rechargeable batteries but this information is still not widely disseminated. (The most-used alkali batteries generally have to go to hazardous waste collection points—although most of us still put them in the trash.) Japan has a special national system for collecting and recycling portable rechargeable batteries, with bins placed at retail and repair shops throughout the country. Sweden also makes battery recycling easy. The hotel where I stayed in the northern Swedish city of Skelleftea had two bins near the door that leads to the parking lot, one for used newspapers, the other for used batteries. The map of that city and other Swedish cities I visited even had "recycling collection points" marked on them with bins for specific items.

option

Televisions are among the most cumbersome electronics to recycle. If you want to recycle a TV in the United States, you have to bring it directly to a recycler, or to a municipal collection point if there is one, or wait for an electronics take-back event. The American divisions of TV manufacturers don't have take-back programs comparable to those available for computers. In Europe, TV recycling works the way computer and other electronics recycling does—or will. In Japan, televisions are collected through the mandated home-appliance recycling program and can be brought to the same collection points as computers.

As for cell phones, the major wireless phone companies—the service providers, that is—do take back phones of all makes for recycling. If the phones are still useable, the phones' memories are wiped and the phones are put into a reuse program. There are also numerous nonprofit donation options for cell phones and computer equipment, but you would be advised to do some research on the organization to make sure what is being donated is actually reaching people who use it. And again, in the United States one has to ask for this information, as it is not made obvious.

▌▎▍▐ ▎

Researchers scrutinizing the environmental efficiency of high-tech equipment have an ongoing debate about the relative ecological merits of recycling and reuse. A study published in September 2004 by CompuMentor, a California-based nonprofit with an interest in reuse, found that refurbishing an old computer for reuse was twenty times more energy efficient than recycling.[2] And by their estimate, only about 2 percent of used PCs find their way to a second-generation user. Eric Williams, who specializes in the life-cycle analysis of high-tech equipment, has calculated that reselling a computer or getting it to a second user saves 8.6 percent of the amount of energy required to make a new computer. Upgrading a computer, Williams estimates, saves 5.2 percent of that amount, and recycling a computer saves 4.3 percent of that energy.[3]

Many manufacturers, including Dell, Hewlett-Packard, Intel, and Microsoft, all participate in programs that facilitate refurbishment and placement of used computers with schools and other nonprofits both at home and abroad, but reuse of high-tech equipment has yet to take off in the United States in a way that would even begin to put a dent in sales of new equipment. And it's almost harder to reuse or pass software on to a second or third user than it is hardware—if you follow the rules of the licensing agreement, that is, but organizations like CompuMentor are working on this and have a program with Microsoft that enables software to be transferred to qualifying second-users. Given the pace of high-tech innovation, until there are substantial design changes that allow new features to be added to older machinery, the refurbished, reused equipment will always lag behind. Although for many users, this may not be a hindrance

There are many admirable programs that make used computer equipment available to schools and nonprofits. Most are aimed at extending the life of relatively new equipment—often equipment from businesses, which tend to replace computers more often than individuals and do so in large numbers. There are nonprofits, like one I've visited in Portland, Oregon, called Free Geek, where donated equipment is refurbished or dismantled for parts that are then used to build new computers that are available either in trade for work at Free Geek or at low cost. And there are programs that send used computers to schools and other nonprofits in less well-off countries.[†]

Most of these programs have inspiring stories that deserve to be told. But what makes these stories interesting and inspiring, I think, has more to do with the social roles they play in education and community development than with solving the fundamental problems associated with e-waste. That said, reuse is one of the best ways to extend the life of high-tech equipment and thereby reduce some of the production and consumption of new products that contributes to the ever-increasing piles of e-waste. Of course, at some point, second-hand computers will reach the end of their useful lives and need to be disposed of as well. Unless we make sure that happens properly—and that design and materials change to make recycling and disposing of high-tech electronics less problematic—we'll have simply slowed or shifted the e-waste stream.

And if any used computer is going to find its way into the recycling system, says Kevin Farnam, manager of environmental strategies and sustainability for Hewlett-Packard's corporate group in Houston, access must be convenient for both consumers and producers. "If it's not convenient, you've lost from the start," says Farnam.[4]

[†] Again one would be well advised to research any donation program carefully, to make sure that all donated equipment is working and going directly to a recipient nonprofit or other qualifying user—and that it doesn't end up being resold, untested, and potentially dumped.

CYBERAGE MINING

If I did send my computer off for recycling, one of the places it might end up is in the electronics recycling facility in Roseville, California, operated by Noranda—a Canadian mining company that's one of the world's major producers of zinc, nickel, and copper—and begun in partnership with Hewlett-Packard, the world's number one manufacturer of printers and their supplies and the world's second largest manufacturer of notebook computers. Why is one of the world's largest metals producers messing around with used computers, machines that contain more plastic than any other material?

"Printed circuit boards are probably the richest ore stream you're ever going to find," said Paul Galbraith of Concurrent Technologies Corporation to the audience of electronics recyclers at the EPR2 in 2002. Mining companies like Noranda and the Swedish copper and zinc giant Boliden Mineral AB, known for extracting metals from the earth, are now mining circuit boards. Scott Pencer of Noranda calls this "above-ground mining."[5] Instead of traveling the globe and making deep holes in the earth to extract their quarry, these companies are shredding and then melting and smelting circuit boards plucked out of old high-tech equipment to extract the valuable metals for resale and reuse. The individual quantities may be considerably smaller but, unlike prospecting for a new lode, a twenty-first-century circuit board miner knows what's going to be found—and the territory is a lot smaller to explore.

The metals of interest to companies like Noranda and Boliden are primarily copper and gold, but circuit boards also contain silver, platinum, and palladium, as well as some other nonprecious metals that can be recycled. Metals generally make up 30 to 50 percent of a circuit board, and while a typical sixty-pound desktop computer is only about 0.0016 percent gold, that gold is almost 100 percent recyclable. As a couple of newsletters I came across put it, "there's gold in them thar circuit boards and handsets!"

The outskirts of Roseville, about a twenty-five-minute drive northeast of Sacramento, are an architecturally unremarkable California locale. New

shopping malls and suburban housing developments are in progress when
I drive through in March of 2004. Surrounded by a chain-link fence, with
some loading docks and parking spaces, and located just off a local highway, the 200,000-square-foot warehouse-like Noranda Recycling facility
gives little, if any, clue from the outside as to what goes on within. From
the area with the front offices where I'm given a hard hat and safety glasses,
I'm guided into a room that seems large enough to house a small regional
airport and is filled with what looks to be acres of computer equipment.
I've entered, I realize, one of the places where these high-tech devices
begin their journey to the electronics afterlife.

Like any highly specialized industry, electronics recycling is a world
unto itself. Wander the floor of an electronics recyclers' conference and
you'll come away with brochures that say things like "Turn Worthless
Waste into a Valuable Commodity," "Certified Destruction," "Crush,
Shred, Pulverize," "From Waste to Raw Material," "Do not waste IT—We
recover it," "Shred-Tech," "Plastic Nation," and "ReCellular." You'll see
ads from companies that make recycling equipment that feature phrases
such as "Whole Lot of Shakin' Going On" and "Separation Anxiety?"
Mark TenBrink, an operations manager for Noranda Recycling calls this
the "tail end of the dog" of the high-tech electronics world.[6]

"Typically we receive about 100,000 pounds a month but can receive
200,000 to 300,000 pounds," my guide Scott Sodenkamp, the operations
manager at the Roseville facility, tells me as we walk the cement floor
among the pallets laden with used computers, printers, fax machines, and
photocopiers. "The most equipment we've ever gotten in at once is 400,000
pounds," he continues. This is just one of Noranda's recycling facilities.
Others are located in San Jose, California, near Nashville, Tennessee, and in
Canada. The Roseville facility, which has about one hundred employees,
receives enough equipment for processing that it usually runs two eight-
hour shifts a day, although it has the capacity to run around the clock.

Some of the recent arrivals at Noranda's Roseville facility are wrapped
in plastic; some are bound with tape. Others are lying in huge cardboard
cartons, and some are stacked on open metal shelves. A few forklifts,
warning lights blinking and gentle horns beeping, are delivering boxes and

pallets of equipment to various workstations. From somewhere out of sight comes the sound of heavy machinery. Despite the fact that this is a destination for discarded equipment, the place is extremely orderly.

Electronics recycling is a relatively new industry in the United States. The International Association of Electronics Recyclers (IAER) calls theirs an "emerging" industry, but a few mining and metals processing companies have been extracting valuable material from used equipment and putting it back into the metals markets for several decades or more. Metals processors like Noranda, TenBrink tells me, began seeing electronics among their incoming scrap in the 1970s, but the finicky business of taking apart complex high-tech electronics equipment and separating out its many materials—so they can be turned into new sources of feedstock—only really got under way in the mid to late 1990s.

According to the IAER, there are from four hundred to five hundred electronics recyclers in the United States. Jerry Powell, editor at *E-Scrap News* had 950 North American e-scrap processors on his database—a number that does not include nonprofits, brokers, or reuse stores. And there are hundreds more high-tech electronics recyclers scattered around the world. This is not insignificant business. In 2003, the U.S. electronics recycling industry was estimated to generate over $700 million annually and expected to grow to four or five times that by 2010.[7] Like companies in any other industry, electronics recyclers run from large and well established to small and struggling. Some are connected to corporations with operations that span the globe; some are fledgling family businesses. And many electronics recyclers specialize in a particular material, type of equipment, or component. To help guarantee a steady stream of material to keep the business running, many such companies have established relationships with high-tech equipment manufacturers. This access is crucial, because fundamental to all electronics recycling is processing enough "raw material" to turn back into a substantial and saleable quantity of feedstock.

Some electronics recycling facilities, like those operated by Noranda and Boliden, are associated with established metals processing companies. "We had a big pile of stuff we didn't know what to do with. They knew what to do but didn't have the stuff," explains Renee St. Denis, manager of Hewlett-Packard's recycling programs, describing how HP and Noranda began working together.[8]

Some electronics recyclers, like Singapore's Citiraya Industries, which had among its customers Intel, Nokia, and Hewlett-Packard, were founded by entrepreneurs who saw e-waste as a business about to boom.[9] Others, like the Finnish company Kuusakoski, receive equipment through the Scandinavian electronics recycling consortium and send disassembled electronics to Boliden for materials recovery. Recyclers elsewhere in Europe and Japan work through similar arrangements with electronics manufacturers collaborating with metals companies, with regulatory and collection systems support provided by their governments.

The U.S. Environmental Protection Agency does not currently license electronics recycling facilities, nor does any other federal, state, or local agency. Some states that have passed e-waste legislation are exploring adding special certification criteria for electronics recyclers to their existing health, safety, and environmental permits. The International Association of Electronics Recyclers has a certification program, but as of mid-2005 only a small number of recyclers had gone through the IAER process. The Institute of Scrap Recycling Industries and other industry groups are developing standards programs, and in 2005, in response to demand from those involved in e-waste issues, the EPA began discussing development of electronics recycling standards.[10]

The e-waste that arrives at the Noranda Recycling facility in Roseville "kind of runs the gamut, from calculators and cell phones to old mainframes," Sodenkamp tells me. But much of what comes in is what he describes as "antiquated materials, old corporate lease equipment, and a lot of over-goods," excess new manufactured product that may have already had a second home. Original equipment manufacturers generate substantial amounts of e-waste, and Hewlett-Packard itself sends a good deal of equipment into Noranda's recycling stream. Noranda also recycles equipment for a number of other manufacturers, including IBM and Sony. Contemplating all these discards, I kept thinking back to what Jim Puckett, director of the Basel

Action Network, told me people said when he asked what happened to their old IT equipment: that it had simply gone "away."

But there is no easy way to make high-tech electronics simply disappear. Historically, high-tech equipment has not been designed with disassembly or materials recovery in mind, so electronics recycling is a relatively slow and painstaking endeavor. The electronics recycling that takes place at a facility like Noranda's entails a detailed process of sorting, then separating out hazardous components like batteries, cathode ray tubes (CRTs), and mercury elements, as well as some plastics for special handling. What remains of the equipment, including circuit boards, is then shredded and later melted and smelted to extract the valuable metals.

When talking about recycling, the phrase "closed loop" is often used. This phrase brings to mind the image of a system without leaks or that of putting a genie back in a bottle. But this metaphor isn't entirely apt, because with a piece of high-tech electronics we're talking about not one but dozens and dozens of genies.

Under current circumstances, it's also relatively unlikely that what comes out of high-tech electronics provides direct feedstock for a new batch of computers, cell phones, or televisions. That said, unless we want to keep extracting and manufacturing recurring quantities of the raw materials that have gone into the world's obsolete high-tech equipment and bury or burn boatloads of high-tech trash, we will almost certainly be seeing more of the kind of recycling that goes on at Noranda's Roseville facility. As Robin Ingenthron of American Retroworks has written, "Even the 'best' mining is usually worse than the 'worst' recycling." But, he says, wherever possible, "Repair and reuse is better than recycling."[11]

Yet, if a machine is merely used, not obsolete, and is still in working condition, it may yet find a new home even if it's shipped to a recycler like Noranda. Equipment that arrives at the Noranda Recycling facility goes into one of two possible recycling streams—what Scott Sodenkamp calls "maximum value recycle" or "destruction and recycle." This, he explains, allows them to do "a real quick triage and decide if there is a second life for the intact equipment or any of its working components."

Everything that comes into the Noranda Recycling facility is scrupulously documented. Each piece of equipment's model and serial numbers are recorded, and each is given an internal ID number. The issue of documenting used high-tech electronics as they enter and flow downstream through the recycling process is a big one. Without the kind of tracking done at Noranda and other legitimate recyclers, this is the point at which equipment can be diverted and loaded onto ships bound for China and other overseas locations for inexpensive, environmentally unsound recycling and/or dumping. "Neither Noranda or HP want to see their company's labels in a heap in a market in a Third World country," says Sodenkamp, alluding to the photographs *Exporting Harm* made notorious. "We do not sell anything to a broker that's not in good working condition. Nothing is sold "as is," he tells me, because "the right thing might not happen then."

An obvious issue in recycling is data security, particularly now that U.S. companies and organizations that collect medical and other personal information are legally obliged to keep it confidential. Noranda's customers can wipe the information off the equipment's hard drive before they send it off or have Noranda do so. Noranda provides customers with certification of data destruction. Some customers, like the military contractor Lawrence Livermore Laboratory, are allowed to observe this part of the process or will ask to be sent a video. "Lawrence Livermore Laboratory usually sends an armed guard," Sodenkamp tells me.

Noranda is not the only company that provides this level of security. This has become standard operating procedure for all reputable electronics recyclers. I observed equivalent procedures at Metech International, at Earth Protection Services in Oregon, and at Kuusakoski's plant in Sweden. Robert Houghton, president of Redemtech Inc., a recycler whose clients include Fortune 500 companies, told me that secure data destruction and downstream accountability were of utmost importance to his customers.[12] Many individuals and small businesses continue to find data security and destruction obstacles to recycling. And it is a serious—yet not insurmountable—concern.‡

---

‡ Data destruction needs to be taken seriously. There is software that wipes hard drives but it must be used properly to be effective.

Once it has been determined that the equipment is going to be disassembled and destroyed for materials recycling, the first step is to check for peripherals and hazardous components. At this point in the process, batteries, cables, power sources, ink and toner cartridges, and bulbs and lighting elements of any kind are removed by hand. They all contain toxic and hazardous substances that should not be shredded along with circuit boards or recycled with plastics or any of the other materials recovered from electronics. Sodenkamp points out the cartons where these hand-separated items are placed. Each will be sent to a different recycler that specializes in batteries, mercury-containing elements, ink and toner cartridges, and so on. CRTs are also segregated, and the copper yokes are pulled from them. All of this disassembly sounds straightforward, but it's time consuming, cumbersome, and requires labor-intensive hand sorting. This is especially true of the older equipment making its way into the waste stream.

"It takes about six to seven minutes to perform the first step in computer disassembly," says Greg Sampson, who works with Earth Protection Services, a national electronics and fluorescent lamp recycler and who gave me a tour of their facility outside of Portland, Oregon, in 2004. "Almost every piece of equipment is different," says Sampson pointing out the manual and electric screwdrivers used for the task. "Sometimes this step takes as much as twenty minutes."

Because Switzerland uses technologically advanced machinery, says Martin Eugster of the Swiss electronics recycling organization SWICO, "It takes approximately 1.3 work hours to process one ton of e-waste in Switzerland while it takes 138 work hours—mostly manual labor—to process a ton of e-waste in India."[13] But most electronics disassembly is astonishingly low-tech. "I think you are familiar with the technologies used in Europe as well as in the U.S. for disassembly—tongs, tweezers, screwdrivers, hammers, heating irons, and ovens," said Dr. Bernd Kopacek, CEO of Eco-efficient Electronics and Services in Vienna, Austria, to the E-Scrap 2004 conference in Minneapolis. "One cell phone may have anywhere from three to twenty-five screws, and dismantling a mobile phone can take between fifteen to one hundred seconds," said Kopacek. Removing the housing from a computer monitor or TV that uses a CRT can take from

six to ten minutes. There are, however, companies developing a special kind
of industrial laser that automates the process of dismantling CRTs and
reduces the potential release of toxics-laced dust. For those companies willing to make the investment, these new tools will come in handy, because
when the WEEE directive is fully up and running, the EU goal is to recycle
75 percent of all the millions of discarded CRTs.

‖ ‖‖ ‖

Desktop computer monitors present a special recycling challenge, but keeping them out of landfills and dumps prevents tons of heavy metals and other
toxics from seeping into the environment. And there are a lot of CRTs waiting to enter the waste stream. While only about 45,000 tons of CRT glass
were recycled in 2000, that volume is expected to be over 1.1 million tons
by 2006.[14] Over 50 percent of a monitor's CRT is glass—much of it leaded
glass. And monitors may also contain barium, copper, silver, lead solder, cadmium, phosphorus, and sometimes tiny amounts of palladium and gold.
Handling CRTs so they don't shatter requires special attention. To compli-
cate matters, CRTs have two layers of glass—one leaded, one unleaded—
and for optimum materials recovery the two need to be separated.

Monitor glass, once separated from the rest of the CRT, is sent for special leaded glass-to-glass or glass-to-lead recycling. Given the limited number of facilities that do this (it involves lead smelting, and with their history of emissions, lead smelters are not popular neighbors), monitor-glass
processing usually happens at locations far from where PCs are dismantled. Noranda's Roseville facility sends leaded glass to a company in
Pennsylvania for the next step in its journey back to usable glass. Earth
Protection Services' has sent extracted CRT glass to the Doe Run smelter
in Missouri and now uses an LG Philips facility in Brazil. While some
researchers are developing ways to more safely deconstruct CRTs, others
are searching for other uses for discarded leaded glass.

Luleå, Sweden, is a city of about eighty thousand on the shores of the
Gulf of Bothnia, where the islands that dot the water were still rimmed
with ice when I flew up from Stockholm the third week in May 2004. I
went to speak with Caisa Sameulsson, a fine-particles metallurgist at the

Luleå University of Technology's Minerals and Metals Recycling Research Centre. One of the first things Samuelsson tells me when I ask about her work is how important it is to "change people's opinion about waste material. People think that nonvirgin material must be dangerous and contain hazardous materials. When you construct something, you have to think that a product should live longer than that one use," she says.

Samuelsson's specialty is the recycling of fine particles and materials that come out of the metals industry—dust, sludges, and slag. She is working on developing a way to use leaded CRT and flat-panel-screen or LCD glass in the copper smelting process. Traditionally, silica or quartz is used as flux— *Option* a substance that helps with the transfer of heat from its source to the metal being smelted. Samuelsson is trying to figure out how these raw materials can be replaced with CRT glass. If this can be done, it would bring about an enormous change in the fate of old computer monitors and TV screens; it would mean that a need for new leaded glass—which is waning with the rise of flat-panel screens—would not be the only way to reuse old CRT glass.

‖‖ ‖‖ ‖

After the monitors and CRTs have been separated from the rest of the computer, the next step, Noranda's Scott Sodenkamp tells me, is to separate out "things that can be tested and resold." This list includes CPUs, CD drives, keyboards, memory cards, working monitors, and fully functioning laptops, printers, scanners, and copiers. These working components and equipment get bundled up and sold via brokers by "nothing less than the pallet load." Where do they go, I ask? "They could go about anywhere," says Sodenkamp.

THE PLASTICS CHALLENGE

Then, what Sodenkamp calls other "low-hanging fruit" gets pulled off the equipment: ferrous metals that can be removed with simple magnets, and aluminum. Compared to the plastics, this is the easy part. "One printer may have four or five different kinds of plastic in it," he tells me as we make our way between workbenches, cardboard boxes of separated computer parts, packing materials, barrels of batteries, and metals drums with

broken bits of equipment. "Plastics are an animal unto itself," says Greg
Sampson of Earth Protection Services. Separating these plastics is a major
challenge in making electronics recycling cost effective. Disposing of plastic in any fashion is costly. Since about 2000, the price to landfill or incinerate plastics ranged from $30 to $110 a ton.[15]

"Plastics is a major issue in our industry," affirms David Weitzman, vice
president of RRT Design and Construction, a Melville, New York, company that designs solid waste processing systems and facilities.[16] As of 2005
most plastics need to be identified by hand, the technical term for which
is "manual resin identification." This means looking for labels or using a
laser probe to determine the type of plastic.

"We have people begging us for material. The challenge is getting
enough postuse material to work with," said Mike Biddle, CEO of MBA
Polymers a California-based company that specializes in recycling the kind
of plastics used in high-tech electronics. To be economically feasible, plastics recycling needs to be done on a very large scale, Biddle explains to a conference full of electronics recyclers and high-tech manufacturers. "We're
beginning to get a demand for recycled plastics from OEMS [original equipment manufacturers] for cost reasons and for green marketing. The problem of getting enough plastics in recyclable form is difficult in the U.S. thus
far, and in North America as a whole, but," he says, "legislation in Japan and
Europe has recently created a larger new source."[17] At an electronics recycling summit hosted by the EPA in early 2005, one of the complaints heard
from electronics recyclers was how few plastics producers are actively
involved in formulating electronics recycling systems and policy.

What MBA Polymers and others have been doing with plastics recovered
from electronics is analogous to what mining companies like Noranda are
doing with metals. With machinery similar to that used to extract and sort
metals, the used plastics are chopped up and separated from the metals. Then
they're ground up into smaller and smaller bits until they become little uniform pellets of plastic—they look like dried peas—that can then be sold to
manufacturers of new products. Yet, when most plastics do find a second
incarnation, it's rarely as part of the same product from which it was
extracted. "I can make plastic flower pots or plastic flowers out of our old ink

cartridge casings," but not new cartridges or printers, at least not right now, Epson's manager of customer programs Shelby Houston told me in 2005.

Packaging of high-tech electronics also uses lots of plastic, and Noranda's Roseville facility has a whole area devoted to sorting packing materials for recycling. In Japan, Sony has been working since 1999 on a way to recycle the polystyrene foam used to pack large televisions, particularly the new LCD-screen TVs. Since 1999 this has been done using limonene, a substance derived from orange and tangerine peel.[18] Limonene dissolves the polystyrene, and the resulting liquid is heated to separate the liquid plastic from the limonene so that both can be used again: the polystyrene as packaging and the limonene as solvent. In 2003 Sony recycled about ninety-nine tons of used polystyrene and turned it into enough new packing for nearly 360,000 large flat-screen TVs.

"Ideally, from the manufacturer's point of view, they'd like to get their own stuff back," says Kevin Farnam, manager of corporate environmental strategies for Hewlett-Packard. "If we get everyone's stuff back there's no incentive to design more easily recyclable stuff. It's a challenge. We need some way to sort by manufacturer. But it's not unachievable. The payoff would be in manufacturers designing more recyclable products with more recyclable content."

CIRCUIT BOARD AS MOTHER LODE

Metal is where the real action is in electronics recycling. To see some state-of-the-art electronics recycling and get a first-hand look at how metals extracted from used high-tech equipment are being processed, I went to Sweden. This may seem a long way to go from the Pacific Northwest, but it's about the same journey that the circuit board from my old computer might make if I sent it to be recycled. After visiting Boliden's Renstrom mine near Skelleftea, I drove back downstream to the port of Skelleftehamn to visit the Rönnskär smelter. That's where dismantled, shredded electronics, disassembled at the adjacent facility run by the Finnish recycling company Kuusakoski are melted down along with newly mined copper ore. This operation, claims Kuusakoski's Web site, is the northernmost industrial recycling plant in the world—and "a major plant even by global standards."[19]

The sky is low and gray over the industrial port area. The temperature is probably in the mid-forties and the wind is blowing in from the water. There is a rail line that parallels the access road leading to the smelter and recycling plant. Some of the Boliden brick buildings date back to the 1930s, but they have an almost Victorian or Edwardian look to them. I can imagine old newsreel footage of the factory with workers emerging with lunch buckets. But what's going on here now is very much twenty-first-century business. For this is one of the only places in the world where high-tech electronics can be both dismantled and turned back into usable industrial material.

To reach Kuusakoski's large metal work buildings and the Boliden plant, I have to stop at a security gate where I'm waved in with a special pass. The entrance to the office area where I'm to meet my tour guide, Ravi Kappel, is up an outdoor metal stairway. To get to the recycling area, we go back downstairs into a large open warehouse area filled with huge metal bins.

Unlike what I saw at Noranda Recycling or Metech International in California, the content of these bins are not limited to computers, printers, and other such high-tech office equipment. And it's clear that most of the equipment has not come directly from manufacturers. These are obviously household cast-offs. The bins here are filled with televisions, computers, stereo equipment, vacuum cleaners, keyboards, lamps, electric mixers, and laptops—anything you can imagine that has a plug—all mixed up together. This is a clear visual reminder that the European Union considers electronics to be anything that uses electricity, while in the United States e-waste is generally considered to be primarily high-tech electronics—equipment with circuit boards, semiconductors, and usually display screens.

I take a closer look into the tall lattice-sided bins. Some items are in their original boxes. There are what look to be stereos from the 1970s and televisions in tacky old cabinetry. There are well-worn desk lamps, parts of video games, electronic musical keyboards, food processors, and hair dryers, even toys. In contrast to the anonymous pale plastic facades of the professional IT equipment that U.S. electronics recyclers receive, this stuff looks very personal. I feel the way I do at garage sales, like I'm getting a glimpse into someone's life.

The items here have been delivered from collection points throughout the north of Sweden, with some equipment coming from northern Norway and Finland as well. The bins of discarded electronics are about seven feet high and look to be at least five feet wide, if not more. There's a lot of stuff here. The Swedish Environmental Protection Agency, called Naturvardsverket, said in 2003 that the "collection of WEEE is continuously high in Sweden. The current annual rate is still about 8 kg. [17.6 pounds] of WEEE per person a year, not including refrigerators and freezers!"[20] After full implementation of the WEEE directive in Europe, all EU countries should be able to make such an estimate. After the loads of discarded electronics arrive, Kappel explains they are sorted and weighed in by type: computers, televisions, lamps, audio equipment, cordless phones, household appliances, and so forth. Kappel shows me how each kind of item is given a bar code and is logged into a computer that weighs up the monthly totals for different types of equipment received.

This is where the regional recycling consortium—here that organization is El-Kretsen—becomes practical reality. The total weight of a particular type of equipment is assessed according to each manufacturer's market share of that equipment. The cost of recycling that amount of equipment is then billed to manufacturers according to their market shares. Having such a coordinated tracking and billing system also helps prevent material from being diverted to illegitimate brokers for export.

After the initial sorting and weighing, the items that are determined to have value as electronic scrap are disassembled. These items are almost entirely equipment with circuit boards that will contain recoverable metals. This equipment is then delivered to a set of six to eight workbenches on an upper level of the facility.

Step one there is manual dismantling with hammers and screwdrivers. All of the workers—about twenty-five in the dismantling, sorting, and receiving area—were on lunch break when we came through, but they had left behind their personal toolboxes, many decorated with the workers' names and playful designs. The workbenches have holes in the middle with chutes that lead down to conveyor belts and sorting bins below. There are bits of broken glass, shattered plastic, and shards of metal scattered on the floor and tabletops, evidence that high-tech electronics are not easy to take apart.

Down below, in big dumpsterlike bins, are plastic computer casings, CRT glass, circuit boards, and different kinds metal, including copper. I ask how the hazardous components, including batteries, get separated. Kappel tells me that this is done by hand. This is also true of processors and contact points that contain beryllium found in some circuit boards. Beryllium is used because it can withstand high heat, but it's also highly toxic. It's a known carcinogen and the dust can cause a chronic disease that is sometimes fatal, so any parts containing beryllium must be pulled out to be recycled separately so they don't go into the shredders. This has to be done by hand and recycling plant workers must learn to spot beryllium components by sight, as the equipment does not come with any consistent materials identification labels.

Scott Sodenkamp had told me that Noranda doesn't accept any beryllium for recycling, but I realize after learning how it's used in circuit boards that what Noranda means is that the company doesn't allow beryllium through its shredders, as these components wouldn't have been removed before arriving at the recycling facility. However, beryllium is also sometimes used as a doping agent in microprocessing, and I wonder if it's possible to extract beryllium when it's bound up with copper in a tiny connector. Boliden's contracts with those who deliver used electronics to their plants carry penalties for the presence of prohibited substances (medical waste and radioactivity) and for restricted substances, including beryllium and mercury, that are not allowed in the smelter.[21]

Sweden has stringent rules about mercury in consumer products and will not allow it to be exported, so used mercury is deposited and stabilized in permanent underground repositories. Because removing the mercury lamps from laptops posed more risk to the workers than it did to send them through the system, they were not separated. I wasn't able to verify this, but Greg Sampson told me that Earth Protection Services removes the mercury elements and recycles them with ballasts from fluorescent lamps. I also asked about tantalum and was told that Boliden doesn't process that. Mobile phones (which contain tantalum capacitors) are so small that neither

Kuusakoski nor the Rönnskär smelter do much with them. Because of their size, cell phones are most efficiently handled by specialized recyclers.

What about plastic and CRT glass? I ask. Kuusakoski doesn't recycle most plastic, Kappel explains, but is looking into it, especially for the hard, usually uniform exterior plastic used for computer housings. He tells me the plastics either get burned with shredded circuit board bits or are land-filled. (Current EPA policy says that if plastics are to be disposed of rather than recycled, it considers "energy recovery"—burning to generate heat used as fuel—preferable to the landfill.) And until a better solution can be found—perhaps like the one Caisa Sameulsson is working on—in Sweden, the CRT glass is sent to a controlled, contained landfill. The zero-waste crowd would not be happy about this, I think.

Kappel then takes me outside where there are huge piles of sorted, dis-mantled electronics. The circuit boards that will be mined for copper and precious and other metals sit in bins awaiting delivery to a shredder I will visit later. The piles of other equipment and parts are enormous. They look like the kind of mounds city snowplows build after a big storm. Some must be six to eight feet high. There are keyboards, computer housings, backs of televisions—all the parts that don't have much recycling value. Under the overcast, chilly Baltic sky, the sight of all this Information Age detritus is par-ticularly bleak. It smacks of dashed hopes for the next new thing, cheap knock-offs, and expensive equipment that turned out to have all the stay-ing power of a plastic fork. As we turn to go inside, I notice a red, white, and blue toy helicopter that has fallen just shy of the rest of the heap.

||| ||| |

After lunch in Boliden's old-fashioned cafeteria—there's pale butter-colored paint on the walls and metal coat hooks, with special ones for guests—Theo Lehner takes me on a tour of the Rönnskär circuit board shredding opera-tion and the smelter. According to the Scandinavian Copper Development Association, Rönnskär is the biggest electronics recycling facility in Europe.[22] All of the material that gets recovered at Rönnskär is metal—copper, gold, silver, nickel, and zinc. Rönnskär recycles about a third of the world's elec-tronic scrap that gets processed in smelters, Lehner tells me.

A metallurgist who's also an environmentalist may seem something of an oxymoron, but I seem to have met two here in the north of Sweden, Caisa Sameulsson and Theo Lehner—even if they might not describe themselves like that. When I first contacted Lehner to arrange a visit, I was given a choice of times to come, depending on what bird migration I might like to see or if I wanted to enjoy the daylight that lingers nearly twenty-four hours a day between May and August. Spring on the northern Baltic coast was chilly and wet, with spots of deep cobalt blue sky peeking between the wind-whipped clouds. Snow was melting up toward the Arctic Circle (only a hundred kilometers away) and the rivers were running high and full. I was struck by the fact that I had traveled to a spot that was about two hours' drive from Lappland to see state-of-the-art electronics recycling.

My tour of the smelter begins in a walled-in yard where tall sloping mounds of circuit boards and circuit board pieces await the next step in their journey back to raw materials. Some piles are full of whole circuit boards as they arrive, just pulled out of equipment at the Kuusakoski plant across the road. Some are made up of big, partially flattened chunks of circuit board—the product of the first round of shredding. Others have smaller, flatter chunks from yet another round through the shredder. The pieces are mostly dark gray and black, with occasional spots of color and shiny metal. They look like parts of a three-gazillion-piece jigsaw puzzle made from a gloomy-hued Jackson Pollack painting. When intact this equipment performed nearly miraculous tasks of information retrieval. Now it's a heap of crunched-up plastic and metal about to be fed into a furnace.

"The scrap business is nothing new for us, but the electronics part is relatively new. But as early as 1967, even before '67, we were doing precious metals recovery. I found a 1967 report," Lehner tells me, "talking about gold recovery from so-called circuit boards. It was 'so called' because they weren't common enough to have a name."

"What is more new," Lehner continues, "is the waste issue. That people are prepared to pay for the processing of waste. The volumes and geography of scrap are also changing." While the electronics that get shredded at Rönnskär come from Scandinavia, the circuit board bits that go into the copper smelter come from around the globe—including some from the Metech plant I visited in Gilroy, California, nearly half a world away.

It's raining hard now and there are puddles everywhere. The Baltic Sea is yards away from where we stand. "Next time you come—come back in a year—and all of this will be enclosed," says Lehner when I ask about runoff and the piles of electronic scrap sitting out in the rain. Lehner tells me that all the water that runs off this area goes to a wastewater treatment plant. The main problem with having the scrap outside, he explains, is the moisture that, especially in the winter, can pose problems during processing.

The chopping and shredding of circuit boards takes place in enormous machines that are big enough to walk inside of. Some—like the one I visited at the Noranda recycling facility in California—have entire control rooms with desks, video monitors, and computers attached. The circuit boards get dumped in one side for initial chopping and emerge on the other side. Then the circuit board bits travel up a conveyor belt into another part of the machine where they are broken up into pieces about half the size of a fist. On subsequent trips through increasingly fine shredding apparatus—rotating metal shears or pulverizing metal balls, depending on the type of shredder—the pieces get smaller and smaller, from about four inches, down to about two inches or the size of toasted wheat cracker, then smaller, to about the size of a U.S. quarter or an average adult thumbnail. This thumbnail size is actually an industry standard known as "a flake."

Some shredders work on a chain-mill system rather than with rotating saws, with big chains at the bottom of a big tub. These work, as I heard a German electronics recycler describe it, "like a hurricane, a coffee mill, or a banana shaker." The aim is to grind, rather than slice, the material as machinery accelerates. Whatever the process, these flake-size shredded bits then go into part of the machine called a "granulator," from which they emerge in pieces fine enough to fit through a mesh screen.

These machines are loud and vibrate raucously. Workers at these shredders—and visitors—wear hard hats, safety glasses, and earplugs or other sound-muffling devices. The one I saw at the Noranda recycling facility runs, I was told, at four hundred horsepower. The shredders I got to see at Noranda, at the Metech recycling facility, and at the Rönnskär smelter had only one or two people working at them: one person to make sure everything ran smoothly from above—aided by video cameras (the Noranda shredder had about six screens for the operator to observe)—and one person below to guide the loading and removal of material. These machines don't separate any metals or other materials at this stage; they just grind and shred.

After the flakes of former circuit boards emerge on conveyor belts, they drop onto a fine screen in another piece of equipment. This machine sifts the small pieces through onto another moving belt where a magnet pulls the ferrous (iron-bearing) metals out into a special collection box. This is known as an eddy current separator. (The fancy new plastics recycling machines that MBA Polymers and others use also work like this.) At this point the ferrous metals are separated from the nonferrous metals (primarily aluminum and copper) and the plastics, which often have been ground up with the copper, gold, and other precious metals. "No one ever delivers hunks of metals to us. It's always contaminated by glass or plastics," says Steve Skurnac of Noranda.[23] At Rönnskär the dust that's captured from all this shredding and grinding is sent to the smelter. Lehner and I watch a load of chopped circuit boards chug up the conveyor belt and into the shredder and see smaller pieces as they emerge on the other end.

We then go inside where Lehner shows me a diagram of how the smelter works. The streams of raw material—ore and scrap—enter from different furnaces for the first step in their cooking process. The scrap furnace used at Rönnskär is a Kaldo furnace, which uses the plastics instead of oil as fuel. Lehner explains that the temperatures are so high in the Kaldo furnace that the plastics are completely incinerated.

Some environmental advocates are adamantly opposed to any burning or incineration. They take this position because no filtration system is emissions perfect and on the grounds that burning—even if the plastic or other waste material is used in place of other fuel—does not put that material to a subsequent solid-state use. They argue that burning is not recycling but downcycling. Measurements done at Rönnskär show that this burning process at least destroys virtually all the brominated flame retardants present in the plastics that go into the smelter. Tests done in

2002 found brominated flame retardant levels below what the European Union considers "detection" limits.

Filters—with scrubbers and a bag-house system—are used to trap any dioxins emitted, emissions that have been previously minimized in a gas purification process. At Rönnskär dioxin emissions were reduced 75 percent between 2001 and 2002. Carbon and lime are used in this process to remove nearly all the remaining mercury. Despite all the filtering, there are still releases to the air of sulfur dioxide, nitrogen oxide, and some metals, including lead, copper, zinc, and some detectable mercury. But all are below what Sweden considers safety levels. And according to Lehner, the sulfur dioxide and carbon dioxide emissions here are almost negligible. "It's the dust that's a problem," with this kind of operation, says Lehner. If you take care of the dust—control and eliminate it—you take care of most of the pollution.

Water released in this initial burning process is filtered and the trapped sludge, which contains copper and other metal particles, is sent back to the smelter for further metals recovery. The remaining water is treated again to remove any remaining particles, which are also returned. But not everything in an operation even as circular as the one at Rönnskär is reused. In 2002, eight thousand metric tons of waste were landfilled on the plant's property. There's no getting around the fact that metals processing is not a squeaky clean business, but this operation is one of the cleanest there is.

Upstairs inside the smelter there's a control room with bay windows. In front of the windows is a bank of computers and control devices with levers that control the machinery, the flow of air, and the materials on the smelter floor. Through the windows I can see large metal hooks ferrying containers and an enormous cauldron. Every now and then there is a clanging and banging as materials—ore and scrap—are deposited somewhere in the system. In the control room, men in blue coveralls and t-shirts are drinking coffee and chatting. One or two are sitting at computer screens. I can't see anyone working on the smelter floor.

"There's a problem when you see someone on the shop floor," says Lehner when I remark how mechanized everything is and how few people are visibly at work. An increase in productivity means doing more with fewer people, Lehner says, and it's obviously much safer in this kind of heavy industry to have fewer people working in these environments. I wonder out loud what the replacement jobs will be. "Bureaucracy," says Lehner. "Researchers and bureaucracy."

Lehner explains that what we're looking at is the pouring of anodes—large molded slabs of metal that will be refined to produce pure copper bars. At the Rönnskär smelter, when the anodes are poured, two metal streams are combined, one that comes from raw ore, the other from electronic scrap—from those chunks and chips of circuit board I saw standing out in the rain-swept yard. I watch the two streams and think: Cyberage meet Industrial Age.

The electronic scrap that's processed here comes not only from local sources in Scandinavia but also from elsewhere in Europe, as well as from South Africa, North America, central Europe, and even the Asia-Pacific region. The amount of electronic scrap recycled and processed at Rönnskär has grown steadily, from about eight thousand metric tons in 1990 to over thirty thousand metric tons in 2002. In 2000 about 30 percent of the copper at Rönnskär came from recycled materials—some of which is electronic scrap—as did 45 percent of the gold and 90 percent of the zinc.

As we talk, the big Kaldo cauldron—it looks positively medieval—begins to tip. "You're lucky," says Lehner, "they're about to make some fireworks." We watch as the boiling tangerine-colored molten copper—it looks like a liquid sun—pours from the giant cauldron. It descends with a great burst of light and flows into something called a converter that burns up the sulfur and, in an oxidation process, separates out the lead, iron, and zinc that are also in the metal mixture. The resulting copper mixture is then poured into molds that make the anodes.

The lead that goes into the copper smelting process with the shredded electronic scrap as a heat control agent is then refined and used again. This is where the potential for using leaded CRT glass comes in. In fact, about half the lead used in the world is recycled, so recycling it from another source, although technologically challenging, is not a conceptual stretch for those who work with metals.

After watching the Kaldo furnace's pyrotechnics, we go through a corridor and down to the floor where the anodes—copper that will be refined further by electrolysis—are made. Here, the molten copper is poured into molds that look like gargantuan rectangular Dutch ovens that circulate on a kind of conveyor wheel as they cool. The copper bars resulting from the electrolysis process are what get sold as processed copper. The slime left after the copper is extracted contains the precious metals—gold, silver, palladium, and platinum—and is sent to the smelter's precious metals division for further separation and refining. I ask if any of the copper processed at Rönnskär goes back into electronics. "We know that some of our copper goes into electrical wire," Lehner says, but given the nature of the metals markets, exactly where this copper goes, he does not know.

Earlier that spring at the Metech International recycling facility in Gilroy, California, I got a good look at how gold and other precious metals are extracted from used electronics. Metech specializes in what the company calls "shred and sample," a process in which small batches of shredded high-tech electronics are assayed for their precious metals contents before being extracted from an entire lot of discarded electronics. Upon arriving at the Metech facility, I was asked to sign a disclosure form—required under California's Proposition 65— acknowledging that was I entering a facility where toxic materials are used, and to walk through a metal detector. The facility's security includes armed guards and twenty-four-hour surveillance. Metech works with precious metals (some are stored on-site) so there is more here of immediate value than large pallets and cartons full of discarded computer and telecommunications equipment. But looks may be deceiving. According to Metech, those bins of used electronics have, on average, concentrations of about twenty troy ounces of gold per ton, two hundred times more concentrated than the amount of gold in ore from a typical mining operation.[24]

Joseph Fulton, Metech's corporate environmental engineer, who is my guide for this visit, tells me that the facility encloses one acre under its roof and employs thirty-five people. A recent survey of electronics recyclers in the United States and Canada found that 70 percent of these companies had less than forty-nine employees and that over half had less than twenty-five. So most fall under the category of small businesses.[25] Metech is among the larger ones.

At Metech, the circuit boards get broken down in machines similar to those I've seen elsewhere. "It's like a big sausage grinder," Fulton says. We watch as a worker, wearing a hard hat and an air-filter mask, shovels the already shredded circuit boards onto the conveyor belt. But what happens next at Metech is something I didn't see at the Rönnskär smelter and that doesn't take place at Noranda's Roseville facility. It's at this point in the operation that I feel like I've taken a big leap, from the quietly rattling keyboards and the cool sounds and sights of the digital universe into Vulcan's flame-belching workshop.

In an open-air area behind Metech's building, a kind of backyard patio, sit a number of roasting tables. The smashed-up little chunks of circuit boards are put on these open roasters, rectangular trays, each about the size of a pool table, sitting over a big burner and under a hooded fan. Nearby, an elaborate pollution-control system with air filters and a bag house to trap dust and other airborne toxics is housed in a contraption that looks like a combination grain elevator and water tower. "California has very strict air pollution regulations," Fulton comments. Still, I can smell what my nose wants to identify as the odor of burning plastic and a whiff of sulfur. I'm surprised that the whole operation isn't somehow enclosed or operated from afar by computer. Fulton explains that samples of the shredded material from a single customer's delivery of discarded electronics are roasted and milled and later analyzed for precious metals contents—gold, silver, platinum, and palladium, but also lesser known metals, ruthenium, rhodium, and iridium.

Part of this process involves melting shred samples in Metech's furnace and turning the molten mixture into "shot examples." I watched as a worker poured liquid metals into small round molds. Some of what remains after this process is sent to Boliden's Rönnskär smelter for further refining. From where we stood at one point, I could feel the heat and see it wobbling off the sides of the furnace.

Fulton tells me that their biggest health and environmental concern is the lead dust that emerges in the shredding and roasting process. Because Metech also recovers gold from industrial waste on the premises—in a process that uses cyanide—many safety precautions need to be taken. Metech's workers are drilled in safety procedures—taking precautions before handling food, leaving all work clothes (these are sent to a special commercial laundry) and shoes on the premises, showering before leaving for the day, and doing what Fulton calls a "self-vacuum." There's also monthly monitoring to test workers' lead levels.

"The direct human health impacts occur mainly during shredding and dismantling," says Martin Eugster of SWICO, "but there are also some indirect impacts—namely air and water pollution which occur as a result of transport, and also as a result of disassembly."[26] It was in Sweden in the 1990s that testing of electronics disassembly workers first discovered polybrominated diphenyl ether flame retardants in human blood—another good argument for designing high-tech equipment with materials that don't contain toxics in the first place.

While the metals markets are highly variable, the reusable metals in circuit boards remain the primary incentive for electronics recycling. This may change if the world resource balance ever tips or if producers are given other incentives—or if there comes a time when recycled plastics are more desirable than what I once heard a retailer call "virgin vinyl." But as I think about this while sitting in a trendy coffee shop where, thanks to my laptop's wireless network card, I have access to libraries full of information, I am once again struck by what Ted Smith of the Silicon Valley Toxics Coalition calls a "clash of the culture of the twenty-first century and a nineteenth-century way of doing things."

"DESIGN FOR THE ENVIRONMENT": TAKING THINGS APART

When Ted Smith says, "We need to internalize costs," or Iza Kruszewska of Greenpeace International and Clean Product Action and Joanna Underwood of INFORM push for manufacturers to take fiscal and physical responsibility for their products throughout the product's life cycle, part of what they're advocating is a concept industrial engineers have come to call "design for the environment." In theory—and in the most ecologically sound of all possible worlds—producer responsibility and designing for the environment are intimately related. Both aim to create products that reduce waste, use fewer resources, and eliminate toxics—and to achieve these ends not just by intercepting a used product before it hits the garbage can.

"We were not thinking holistically," said Pat Nathan, Dell's sustainable business director at the E-Scrap 2003 conference, reflecting on what she called the industry's "evolving view" of product stewardship. "I don't think the whole IT industry was thinking that way," commented Nathan reflecting on the last five to ten years. "People were begging for an extra PC. Now they're looking at how to get rid of it."

If high-tech electronics are hard to take apart, you don't know exactly what's in them, and some of their materials are toxic, they'll be difficult to recycle. And if it's assumed that used high-tech products are simply going to be trashed regardless of their impact on the environment and human health, there's little incentive for manufacturers to make products that facilitate materials recovery and reuse. But designing for the environment involves more than thinking about recycling. It means minimizing the use of resources and reducing, or better yet, eliminating toxics and waste and detrimental impacts throughout the production process. It also means developing production methods that don't pose safety risks for workers. Ideally, the ecological footprint of a product designed for the environment would be almost imperceptible.

Virtually all major high-tech manufacturers now have design for the environment programs. Even Microsoft, which makes mostly software, has a note in its corporate responsibility report—albeit not detailed—about materials used in its products. "HP's design for the environment program has been in place since the early 1990s," David Isaacs, Hewlett-Packard's director of global public policy, told me in 2003. This timing coincides with the start of other major high-tech manufacturers' similar programs. As Isaacs and his peers at companies like Dell and IBM explain, making computers and all their companion gadgets easier to dismantle—using snap-in parts instead of screws, eliminating glues and adhesives, reducing the number of plastics used, making ink and toner cartridges refillable and recyclable—are all part of designing for the environment.

Designing for the environment also means improving a product's energy efficiency, creating a product whose life span can be extended, and making products safer and more comfortable to use—call this ergonomic correctness. Reducing packaging and using recyclable and nontoxic packing materials is part of this kind of design priority too, as is providing easy access to information about product design, materials, and recycling options. Another way of reducing materials and use of resources is to make equipment smaller and lighter and to reduce the number of machines and accessories. For example, eliminating the need for three machines by creating an all-in-one fax, printer, and scanner, or substituting flat screens for CRT monitors, or creating internal CD/DVD drives in place of separate devices—are all evolutions that fall under design for the environment heading. Panasonic's Web site shows how, between 1980 and 2000, it has reduced the number of parts in a television to facilitate recycling, in part by compressing internal components into a circuit board.[27] Such miniaturization and simplification is prevalent throughout the high-tech industry. Dell, somewhat hyperbolically, calls it "immaterialization."[28]

Designing for the environment also involves labeling parts by material and listing what are, in effect, a product's ingredients, something that seems simple, sensible, and uncontroversial. Knowing what something is made of is a prerequisite for recycling. But when labels alert consumers to materials they want to avoid (or think they should avoid)—as they have with food, pharmaceutical, and tobacco products—they often become controversial. Materials listing also becomes tricky when that information is considered a trade secret by the manufacturers in question, as it often is with high-tech electronics components.

In contrast to other things that corporations do to enhance their "greenness"—like giving to charities, participating in community clean-ups, and improving in-office resource practices, all of which are extremely worthwhile—changing product design requires an additional level of commitment. For example, if a company installs an ecoroof on one of its factories but continues to increase profit margins by producing ever-greater numbers of ecologically unsustainable products, it would seem fair to ask if the company has tackled the hard part of reducing its contribution to adverse environmental impacts. With growing scrutiny of manufacturers' environmental performance and increasing concern over the impacts of toxics in high-tech electronics, more and more companies realize that it's in their long-term interest to deal with end-of-product-life issues before those products hit the assembly line.

▌█ ▌▌▌ ▌

Although American high-tech manufacturers may resist this characterization, to a large extent many of the design changes in high-tech electronics have been prompted by European regulations and design standards. Hewlett-Packard's design for the environment program was "in place long before WEEE or RoHS were enacted or in place," Hewlett-Packard's David Isaacs tells me. But he says, these directives will have an impact, because "as they become legal their requirements will become inescapable."

The RoHS directive, which becomes effective in 2006, will require the elimination—with certain exceptions—of lead, mercury, cadmium, hexavalent chromium, and two types of flame retardants (polybrominated biphenyls and certain polybrominated diphenyl ethers) from all new electrical equipment sold in Europe. Among the changes these restrictions will bring are the end of the lead solder typically used in circuit board assembly, the end of cadmium in portable batteries, of cadmium and chromium in paints and inks, of mercury switches, and of the penta- and octabromodiphenyl ethers used to render certain plastics (and upholstery foam) fire resistant.

Just to be clear, RoHS does not mean the end of leaded glass in CRTs, of the mercury lamps used in flat-panel display screens, of beryllium processors, or of polyvinyl chloride–encased wires and cables. It also does not require the elimination of the deca-BDE flame retardant—the PBDE most commonly used in the hard plastics that house electronics—nor of the tetrabromobisphenol A (TBBPA) flame retardant used in nearly all circuit boards and many other plastic parts of high-tech equipment. It does, however, require that these and other exceptions to the hazardous substance restrictions be revisited every four years with a view to substituting materials determined to be less toxic, leaving open the possibility that more substances could be added and standing exemptions removed. In language that no doubt makes American free market policy proponents nervous, the RoHS directive asks that the precautionary principle be applied in assessing the safety of listed toxics in applications initially exempt from RoHs and in evaluating substitutes for these hazardous substances.

Because RoHS applies to all electronics sold in the European Union regardless of where this equipment is made or the manufacturer's home country, U.S. companies—including Dell, Hewlett-Packard, IBM, Intel, and Apple—will be making products to meet EU requirements that will be sold worldwide, as will Japanese manufacturers. Chinese high-tech companies, which do a huge export business to Europe and make components for many American and European companies, are also scrambling to comply with RoHS and WEEE requirements. It's simply not practical or profitable for high-tech manufacturers with factories in every time zone—that turn out products also sold in every time zone—to design one model that can be sold in Denmark and another that can be sold in Detroit. "EU design requirements will become global requirements," says Isaacs.

Regulations are not the only European influence on high-tech product design. Since the 1990s, guidelines known as "ecolabels" formulated in Europe—especially Sweden's TCO standards and Germany's Blue Angel program—have quietly been nudging high-tech companies like Dell and Hewlett-Packard toward making their products more environmentally friendly. Ecolabels like Blue Angel, TCO, Japan's PC Green Label, Scandinavia's IT Eco Declaration, and more specifically focused labels like Energy Star (which rates energy efficiency)—used in the United States, Australia, EU countries, Japan, and Korea—and Canada's Environmental Choice (which rates energy efficiency, air emissions, and CFC use) have increasingly become factors that decide a large purchase order. For example, the Clinton administration issued a directive specifying that the federal government—probably the country's largest single purchaser of high-tech electronics—would buy only Energy Star–certified products. Hewlett-Packard reports that in 2004 purchase orders that included environmental criteria were up by 95 percent from 2003 and 660 percent since 2002.[29]

Yet while all electronics sold in the European Union will be RoHS compliant, all design for the environment features are not present in all products made by any one manufacturer. For example, not all ink cartridges are recyclable, and one manufacturer may recycle its toner but not its ink cartridges, or may recycle some but not all models of cartridges. And there's a wide variation of environmental efficiencies within any suite of products offered by just about all high-tech manufacturers, which means consumers who want these features must do research on a product-by-product basis.

"No simple label or seal of approval will address all consumer concerns," says Hewlett-Packard's David Isaacs. Were the EU requirements a positive development? "Yes," if they help "improve the design of products with an eye toward environmental protection." But, he cautioned, it's important not to "get caught up in slogans or banning certain substances because it sounds like the right thing to do."

The task of ensuring that all parts of a particular piece of equipment meet RoHS standards—or any other environmental criteria—is complicated by the fact that most manufacturers don't make all their own products' components. For example, in the 2005 iteration of its "design for the environment" information, Dell says that it's working with over 140 different suppliers. Sony's Web site mentions auditing "about 4,000 suppliers."[30] So when original equipment manufacturers talk about the "supply chain," they're talking about dozens and dozens of different companies located all over the world—many of which make components for more than one manufacturer. The upside of the extended geography of high-tech manufacturing is that it enhances the likelihood that high environmental standards will be adopted more widely than they would otherwise. But this global scatter also means diligent oversight is required.

Which prompts the question: How will anyone know if manufacturers are meeting the materials and design standards? There is, of course, the honor system, but there are also penalties for noncompliance with RoHS and WEEE (penalties "to be determined," say the directives). The WEEE directive also requires manufacturers to provide recyclers with a materials list and encourages the use of recycled materials. This list will not detail every chemical that goes into making a semiconductor—into each plastic or every silicon wafer, for example—nor will WEEE require disclosure of proprietary information. But such a list will tell a recycler if a computer contains anything that requires special treatment and will help identify plastics.

American high-tech companies often say that WEEE will not influence what they do at home because national electronics recycling is not mandated in the United States. Yet all U.S. electronics manufacturers who sell their products in Europe will have materials lists ready for recyclers, and the same products these companies sell in Berlin, Oslo, Barcelona, and Tokyo will also be sold in Boston and Dallas. It would seem a terrible error in customer relations if the U.S. branch of a major high-tech manufacturer refused to disclose a materials list prepared under WEEE to a recycler in California when that same list has been made available in Sweden.

A curious gap in this upward trend in environmental standards may be equipment made by Chinese manufacturers that could—as things stand in 2005—be sold in the United States. Although in early 2005 China announced regulations that will mimic the RoHS standards, many Chinese companies are worried about their ability to meet such standards. In what could be something of a great environmental leap backward, unless U.S. regulations change, it would be possible for Chinese high-tech equipment that is not RoHS compliant to be sold in the United States. But with increasing state regulation that mimics EU standards, such exports will become more problematic—and such equipment is likely to create headaches for recyclers.

GETTING THE LEAD—AND OTHER TOXICS—OUT

With regulation of certain substances on the horizon, some manufacturers have already altered product design to eliminate certain toxics. Apple, Dell, Ericsson, Hewlett-Packard, IBM, Intel, Philips, and Sony are among the high-tech manufacturers that have also eliminated PBDEs from their products—specifically from plastic parts weighing more than 25 grams (one ounce equals 28.35 grams). In response to the TCO and Blue Angel ecolabels, Dell, Hewlett-Packard, IBM, Apple, and other manufacturers are eliminating *all* halogenated flame retardants from plastic parts that weigh over 25 grams and are investigating the possibility of using halogen-free circuit boards. Apple has already stopped using the TBBPA flame retardant in parts heavier than 25 grams.

Eliminating lead solder poses some challenges—among them temperature and ease of manipulation. "One of the problems with lead-free [solder] is that the melting point goes up and affects all the other materials," explained Richard Puckett of Pixelworks, a company that designs and manufactures semiconductors for high-resolution visual applications like high-definition TVs. Lead-free solder—tin and silver or copper is one combination—melts at higher temperatures than lead, so other circuit board components must be engineered to withstand greater heat. The trick, he told me in April 2004, is "to find a substitute that will work in a similar way."

"We can't eliminate all application of lead, but it will be good to eliminate many of them," says Timothy Mann, IBM's program manager for environmental policy.[31] A number of high-tech manufacturers and industry observers have questioned whether focusing on lead in circuit boards is the best use of resources in terms of "greening" high-tech products, as the transition will be expensive. But as a representative of the capacitor manufacturer Kemet told the *Toronto Star* in 2004, "The debate as to whether it's a smart idea is over. The legislation is passed. We have to do it. So now we have to figure out how to do it."[32] Despite the short timeline, manufacturers are preparing to meet the 2006 lead-free deadline set by RoHS, and some have already eliminated lead solder from a number of products.

||||| ||| |

High-tech manufacturers say that their design for the environment programs have been undertaken voluntarily and are not the result of legislated mandates. But it's unlikely that without outside pressure—whether from regulation, consumers, scientific discovery, or NGOs—that high-tech manufacturers would be confronting what goes into—and comes out of—their products as they have begun to do. "Eventually this waste affects all of us," Sheila Davis, then with the nonprofit Materials for the Future Foundation, said in 2002.[33]

WHAT'S FAIR?

Not all electronics recycling in the United States takes place in open workplaces like the Noranda and Metech facilities I visited. A great deal of electronics are dismantled for materials recovery by U.S. federal, state, and county prison industries—a fact that poses serious questions about labor practices and business competition. In late 2004 Leroy Smith, a Bureau of Prisons employee for over a decade, filed a formal complaint with the U.S. Occupational Safety and Health Administration (OSHA), claiming that workers at the Atwater Federal Penitentiary computer recycling facility run by the Federal Prison Industries have been exposed to heavy metals dust. In June 2005 I went to visit Smith, and to learn more about this cheap labor being used for computer recycling.

On the eastern horizon, beyond the dairy farms and orchards of California's San Joaquin Valley, the Sierra Nevadas are just visible through the haze. To the west of the two-lane road, where I've pulled off about ten miles north of Merced, sits the Atwater Federal Penitentiary, its tower and low-slung buildings the same mustard yellow as the dry fields that stretch beyond the chain-link fence and concertina wire. At this maximum-security prison inmates have been smashing computer monitors with hammers, releasing dust that contains lead, cadmium, and barium, as well as other toxic substances. These inmates are employed by the electronics recycling business of the Federal Prison Industries (UNICOR). With sales that have nearly tripled between 2002 and 2005, electronics recycling—a service UNICOR began offering in 1994—is the company's fastest growing business. But this work, says Leroy Smith, Atwater's former safety manager who is on leave under whistleblower protection, is being done under conditions that pose hazards to the health of prison staff and inmates—conditions that would not be tolerated or allowed in the private sector.[34]

According to Smith, workers at Atwater's UNICOR computer recycling facility are routinely exposed to lead, barium, beryllium, and cadmium, and use safety equipment that doesn't meet OSHA standards. Neither staff nor inmates, says Smith, were properly informed about these hazards. After his superiors sent OSHA a report that downplayed and denied these problems, Smith sought whistleblower protection.

With revenue of ten million dollars in 2004, seven locations—Atwater, California; Elkton, Ohio; Marianna, Florida; Tucson, Arizona; Lewisburg, Pennsylvania; Fort Dix, New Jersey; and Texarkana, Texas—and roughly one thousand inmate employees who in 2004 processed nearly forty-four million pounds of electronic equipment, UNICOR is one of the country's largest electronics recyclers, and its prices are tough to beat.[35]

"UNICOR's program is labor-intensive, so capital machinery and equipment expenses are minimized, this helps keep prices low," says UNICOR.[36] With a captive workforce, UNICOR's electronics recycling program can afford to be labor intensive. Unlike the private sector, UNICOR does not have to pay minimum wages—wages in 2003 were $0.20 to $1.26 an hour[37]—or provide benefits. Established by the federal government in the 1930s, UNICOR's work programs are voluntary, and inmates earn money to pay child support, court-ordered fines, and victim restitution. A for-profit corporation run by the Bureau of Prisons, UNICOR is not taxpayer supported. But its pay scale would not be possible without public support of the inmates it employs.

In 2004, UNICOR's electronics recycling facility at the Lewisburg, Pennsylvania, federal correctional facility offered the Pennsylvania Department of Environmental Protection a per-pound processing quote that was one-quarter the price quoted by private electronics recyclers: UNICOR's price was $0.05 per pound, the private recyclers were asking $0.20 per pound.[38] The contract went to UNICOR.

In early 2004 Andy Niles, vice president of Scientific Recycling, an electronics recycler in Holmen, Wisconsin, lost a county recycling contract to Badger State Industries, the Wisconsin state prison industry. This caused Niles to lay off six employees—about 25 percent of his workforce. "I welcome the competition, but let's level the playing field," Niles told me in June 2005.

▌▐▎▐

"The Recycling business group provides both government and non-government customers an environmentally friendly way to dispose of electronic equipment," says the UNICOR Web site.[39] But Smith and staff members at other prisons where UNICOR recycles computers—as well as inmates—say otherwise.

Instead of investing in state-of-the-art disassembly equipment, UNICOR distributed ball-peen hammers and cloth gloves to inmates working at Atwater. "The gloves ripped easily and there were lots of bad scratches and cuts," recalls an inmate who worked there in 2002.[40] Staff and inmates who worked at UNICOR's Elkton and Texarkana operations have similar accounts of broken glass, noxious dust, and injuries resulting from inadequate tools.

"Initially, there was not a whole lot of training," Smith told me. "We began electronics recycling in April 2002. We started with nothing. UNI-COR gave us no procedures for dismantling CRTs. There wasn't a whole lot of guidance from them. We had improper tools and the staff was not properly trained to train inmates. Because of the lack of training there was a mass of injuries at the beginning."

"We were given light-particle dust masks and the stuff would get in behind them. In the glass breaking room guys would be pulling junk out of their hair and eyebrows. We were coughing up and blowing out all sorts of nasty stuff and open wounds were not healing," the former UNICOR computer recycling worker told me. He also described how the coveralls inmates wore on the job—kept on during breaks and meals—would come back from laundering with glass and metal dust in rolled cuffs. Work boots were worn outside the factory, potentially contaminating other areas, an occurrence OSHA regulations are designed to avoid.[41] Prison staff, say Smith and others, wear regular uniforms and shoes, allowing contaminated dust to be transferred to personal vehicles, homes, and family. I thought about this when Smith's small granddaughter wandered into our interview at his home, wearing only a diaper and a t-shirt.

Air samples taken at Atwater in 2002 found lead levels two, five, and seven times OSHA's permissible exposure level and cadmium nearly eight and thirty times the OSHA standard. Wipe samples taken in 2002 found lead, cadmium, and beryllium on work surfaces and inmates' skin. Blood and urine testing also found barium, cadmium, and lead, some at elevated levels.[42]

UNICOR's computer disassembly process releases so much lead that its dust qualifies as hazardous waste.[43] Smith and former staff at UNICOR's Elkton, Ohio, facility say this waste has been improperly handled. "Prison staff were removing the filters that collect the dust from the glass breaking without wearing respirators and putting these filters in the general prison trash," Smith told me. Documents dated 2005 and made available by Public Employees for Environmental Responsibility confirm these accounts. Leroy Smith showed me photographs of work tables covered with thick layers of pale gray dust.

So how toxic is this dust? The lead found in Atwater's air in 2002 would eventually produce elevated blood lead levels, explained a leading environmental and occupational health scientist.[44] Over time, this lead lodges in bones and can cause kidney or neurological damage. While levels found at Atwater would not cause acute lead poisoning, I was told, they are definitely unhealthy and would likely have adverse effects.

Inhalation exposure is the most dangerous in terms of getting lead into the body, says Howard Hu, professor of Occupational and Environmental Medicine at the Harvard School of Public Health.[45] "If wipe samples show elevated levels of lead there's serious potential for hand contamination, especially if workers eat on the job or smoke cigarettes," Hu told me. What makes exposure at the upper limits of OSHA standards worrisome, explained Hu, is that current science indicates that these levels—and EPA's permissible levels—are likely far higher than what is truly safe, especially where children are concerned.[§]

"Busting up monitors exposes you to a lot more risk. But broken monitors saves on shipping costs," says Greg Sampson of Earth Protection Services. "Broken you can fit about one hundred into a carton, whereas only thirty-five or so will fit if they're intact. We don't break ours up." At Atwater, Leroy Smith told me, broken CRTs are packed in cardboard cartons and sealed with plastic wrap. While Noranda, Earth Protection Systems, and other private recyclers told me where they send CRT glass for recycling, UNICOR would not, saying that this information—and the destination of any other demanufactured material—is "proprietary."

▌▌▐▐▌▐

One of the inequities of competing with UNICOR, say private recyclers, is that because UNICOR works behind bars it doesn't have to be prepared for unannounced OSHA inspections. OSHA officials told me that its inspections of UNICOR computer recycling facilities—none unannounced—had all

---

§ "Everything we know seems to indicate that there is no threshold level of lead exposure for adverse consequences on children's intellectual ability. Which gets translated as: there is no safe level," Dr. Bruce Lanphear, director of the Children's Environmental Health Center at Cincinnati Children's Hospital Medical Center and author of recently published studies assessing the effects of lead on children's neurological development, told me in July 2005.

taken place in 2004 and 2005, despite the fact that some factories opened some ten years earlier. Security concerns are also responsible for some of the working conditions that wouldn't be accepted in the private sector. For example, Jeff Ruch of PEER told me, "Due to security concerns it is inefficient to move workers back and forth from work areas to other areas during the work day. There are also concerns about smuggling sharp objects out of work areas, so they have had food service in contaminated work areas."

Following the 2002 air tests that revealed elevated levels of heavy metals, UNICOR suspended CRT breaking at Atwater for several months and brought in new safety equipment. Subsequent air testing in December 2004 found lead, barium, and cadmium but at amounts below OSHA's action levels. But these samples were not taken around inmates "involved in the deliberate breaking of monitors," says the OSHA report.[46] Barium, beryllium, cadmium, and lead were also then found on work surfaces, and barium, cadmium, and lead showed up in the workers' dining area. Having these substances in both work and eating areas, the OSHA inspector wrote, "could pose a cross-contamination potential to workers through ingestion."[47] Yet these findings don't violate OSHA standards because, as the OSHA inspector noted, there are "no standards or regulatory levels for these metals on surfaces." In May 2005, OSHA said it was "unable to substantiate" any of these conditions.[48] OSHA inspections, however, deal only with what's found during a particular visit.

Much of what UNICOR recycles comes from the federal government, which buys about 7 percent of the world's computers and recycles at least half a million each year, according to the EPA. In 2002, the Department of Defense sent some seventeen million pounds of used electronics to UNICOR for recycling.[49] UNICOR's Web site also lists as clients the University of Maryland, University of Connecticut at Storrs, Michigan State, Penn State, and other universities.[50]

UNICOR requires the processors it works with to certify that all material is resold, reprocessed, or repaired for reuse and is not landfilled or exported for dumping. According to a Bureau of Prisons spokesperson, all disassembled components go "to permitted, certified recyclers and processors." However, there's no government certification program for electronics recyclers,[51] so "permitted" and "certified" simply means complying with standard business regulations.

UNICOR's no-export policy prohibits shipping material to any party that cannot legally receive U.S. products or technology the government deems "sensitive." The list of prohibited parties, says the Bureau of Prisons, is maintained by the U.S. state and commerce departments. The countries on that list are Cuba, Iran, Iraq, North Korea, Libya, Sudan, and Syria. In reality, this policy can't prevent export of e-waste to places like China or India, as there is no U.S. law that prohibits the export of e-waste if it's destined for reuse, repair, or recycling.

▌▐▐▌▌

"We can do without it but are we willing to do without it?" asks Craig Lorch of Total Reclaim, whose company is among the thirty or more that have pledged not to use prison labor.[52] The United States, it seems, is the only industrialized nation that uses prison labor for electronics recycling. Following pressure from Texas Campaign for the Environment, Silicon Valley Toxics Coalition, and the other members of the Computer TakeBack Campaign, Dell canceled its contract with UNICOR, as did the state of California, and Hewlett Packard pledged not to use prison labor for recycling. But as of mid-2005, neither the U.S. Electronics and Manufacturers Association nor the IAER had taken an official position on prison labor. Meanwhile, in midsummer of 2005 as part of its response to Leroy Smith's complaints, the Bureau of Prisons released a letter admitting that staff and inmates at several UNICOR computer recycling facilities were routinely exposed to unsafe levels of heavy metals.[53] When I met with Smith he was determined to pursue all the options available to him through the Office of Special Counsel, including a request for a congressional hearing. As I write, Smith's case continues to work its way through the Department of Justice.

However Smith's case turns out, it seems clear that relying on workers who aren't paid a living wage and who work in unhealthy and environmentally unsound conditions displaces rather than solves the e-waste problem. And without transparency and accountability from recyclers, it's impossible to know how this material is treated and where it goes.

CHAPTER NINE

A Land Ethic for the Digital Age

The most difficult of all our retreats will take place in the war we have been waging against our biosphere since the Industrial Revolution.[1]
—Hans Magnus Enzensberger, *The Hero as Demolition Man*, 1989

The public must decide whether it wishes to continue on the present road, and it can do so only when in full possession of the facts.[2]
—Rachel Carson, *Silent Spring*, 1962

No matter what amazing innovations the next decades of high tech bring, one thing is certain: e-waste is not going to disappear, neither is the prodigious use of chemicals in high-tech manufacturing, nor will we see an end to all natural-resource extraction. Computers, cell phones, and the whole universe of digital devices may become smaller, more powerful, and more efficient, but there will be more of them than ever. Microchip circuitry may be as invisible as the network of nerves on a dragonfly's wing, and whole libraries may appear on our desktop screens apparently out of thin air, but unless some radical changes are made in the way we design and produce our Information Age gadgetry, its ecological footprint will never really be reduced.

By the end of 2004, there were over 820 million computers in use worldwide. That year, sales jumped between 11 and 15 percent—a growth rate over ten times faster than that of the world's human population.[3] By June 2005, the Computer Industry Almanac recorded some 898 million

PCs in use around the world, with laptops accounting for nearly 230 million of those computers, up from 31 million ten years previously.[4] At this rate, in 2007 there will be over 1 billion computers—about one computer for every six people on earth.[5]

Worldwide sales of semiconductors and microprocessors were, despite ups and downs in the market, both over 10 percent higher in February 2005 than they were at that time the year before.[6] More semiconductors and microprocessors mean more PCs, cell phones, DVD and MP3 players, televisions, digital cameras, and other high-tech goodies. "Technology has made it possible to process more silicon for less money, which is a benefit to consumers," John Greenagel of the Semiconductor Industry Association told me in May 2005. As more new high-tech electronics arrive in homes and offices, more existing equipment will be discarded as obsolete or defunct. And as high-tech equipment reaches more and more people, e-waste will become an issue that virtually no country can afford to ignore.

The United States—which has more computers than any other country—is not ignoring e-waste, but by the end of 2005 it has not even sketched out a national system for dealing with its high-tech trash. By the middle of 2005 the United States had about 230 million computers in use, over three times as many as Japan, the world's second most PC-populous country. But use of computers is growing so quickly in China, the third heaviest user of PCs, that its computer numbers are expected to exceed Japan's in 2007.[7] Sales of PCs are also growing quickly in other large, geographically diverse, and heavily populated countries, among them Russia, Brazil, and India.[8]

On desks from Bhutan to Micronesia, words, images, and sounds emerge with the tiny gesture of a finger, thanks to silicon mined in China, Norway, or perhaps Canada, that was crystallized in Germany, polished and sliced in Oregon, and then etched into microchips in Arizona, Costa Rica, or Ireland. Popped into circuit boards in Malaysia, Chengdu, or perhaps Korea, Romania, or Vermont, these microprocessors run on electrical charges conducted by copper refined in Sweden, gold from Indonesia, and stored by a speck of tantalum from Australia, or maybe Rwanda, that was processed in Pennsylvania or the Netherlands. They are surrounded

by plastics made flame resistant with bromine from Arkansas—as specified by designs dreamed up in California, Texas, Tokyo, Finland, or
Bangalore—and turned out in factories in Mexico, Singapore, or Scotland.
And when they can no longer be made to work, our computer equipment
begins its circuitous return journey to smelters, refineries, and plastics factories, with a distressingly large percentage of our electronic discards—
about 90 percent as of 2005—slowly degrading in landfills or being liqui-
dated in municipal incinerators.

Meanwhile, the legacy of the high-tech industry's use of toxic chemicals persists in groundwater and soil, the vapor seeping into private
homes, and in the bodies of some of those who have worked with these
chemicals during their manufacture or disposal. With the right sort of systems to collect and process this materially complex machinery, the gap
between electronics recycled and trashed could be narrowed, especially
if accompanied by changes in design and materials.

As Europe begins to implement its WEEE directive, requiring recycling of
used high-tech and other electrical equipment, citizens in Japan are duti-
fully separating their garbage into a list of categories long enough to rival
choices on a sushi bar menu and are depositing obsolete electronics for
recycling pickup at the local post office. China is working on national regulations requiring manufacturers to participate in the recycling of used
electronics. But in the United States, where there are over seventy-six computers for every hundred people,[9] and we have the potential to generate
more e-waste than any other country, we continue to struggle through an
often confusing array of options guided by little consistent or explicit public information.

In 2005, Congress and about two dozen legislatures introduced e-waste
bills. Thus far, at least thirty-two states have considered e-waste legislation,
and a dozen states have passed bills or other legislative measures concerning e-waste. Legislation that mirrors Maryland's bill, which requires computer manufacturers to establish their own recycling program or participate
financially in a state-run program, has been introduced in Massachusetts,

New Jersey, and Wisconsin. New York City is considering a similar bill, and ten Northeast states are working together to draft e-waste legislation that would be introduced throughout the region.

In February 2005 the Competitive Enterprise Institute, a Washington, DC, nonprofit "dedicated to the principles of free enterprise and limited government," released a report on electronics recycling. It claimed that "a single 120-foot-deep, 44-square-mile landfill could accommodate the United States' garbage for the next 1,000 years," and that modern landfills can safely handle e-waste.[10] But many state legislators seem not to have fallen for this message. Despite the George W. Bush administration's federal policies—and preferences—that consider such regulations a burden to business and an obstacle to economic growth, some state governments are willing to challenge the status quo. This is especially true when the disposal of potentially hazardous waste is involved, since local communities are responsible for managing such waste.

Still, any comprehensive system for electronics recycling in the United States seems a long way off. The upshot is that most American households are still stashing their e-waste in closets, basements, garages, and attics—and it's still perfectly legal in most communities to toss an old computer in with the rest of the trash.

However helpful they may be, nationally coordinated recycling programs alone will not decrease the amount of old, obsolete, and discarded electronics in years to come. To do so will require design changes in hardware and software that substantially extend equipment's life. But recycling programs will make responsible disposal easier. Increasing recycling should also put more postconsumer materials back into production and get more working equipment to second users. It will also mean less smashed leaded glass, fewer banged-up circuit boards, and fewer loads of plastic malingering in landfills where they can leach toxics into the environment.

If accompanied by sufficient oversight at ports and loading docks, putting the WEEE directive to work should also mean fewer containers of waste electronics being shipped abroad from Europe for low-cost dismantling in southern China and other parts of the world hungry enough to relinquish their environmental and personal health for this labor-intensive work. But nothing is going to change overnight. Until high-tech electronics are made easier to disassemble and contain fewer hazardous materials, e-waste will undoubtedly continue to be exported from countries with high wages and stringently enforced environmental and labor standards to places like Guiyu for inexpensive, environmentally unsound recycling.

With ongoing international and scientific scrutiny of environmental damage wrought by the rudimentary processing of e-waste, the Chinese government has begun to regulate the e-scrap industry and is touting the improvements being made in Guangzhou. Yet according to a 2005 report, women there continue to dismantle circuit boards and CRTs in home workshops while their small children stand nearby. Open burning and sulfuric acid baths have officially been banned in Guiyu, but have sometimes been replaced by manual labor. Instead of melting plastic to expose copper wires, workers are peeling the coating off by hand.[11] New municipal water and sewage treatment plants are being built and the government wants to turn Guiyu into what the Xinhua News Agency called "a national showcase center for recycling discarded electronic and information technology products."[12] But I wonder how difficult and expensive it will be to clean up the hazardous waste that has accumulated around Guiyu and in the Lianjiang River. I also wonder, as new dumps of high-tech trash turn up—as they have in Lagos, Nigeria, were some five hundred container loads of e-waste have been arriving every month, much of it American in origin and much of which ends up burnt and smashed—what kind of vigilance will be required to ensure that new Guiyus do not emerge elsewhere.

▌▐║▐║▐

In early 2005 the U.S. Environmental Protection Agency held an electronics recycling summit. Among the issues participants grappled with—and on which there is no industrywide or national policy—are that of certifying electronics recyclers, exporting electronic waste, and of using prison labor for electronics recycling. There were complaints voiced about the difficulty of dealing with products designed in ways and with materials—

specifically plastics—that make recycling complicated and therefore pro-
hibitively expensive. But loudest of all complaints from the assembled
e-waste professionals was that the United States has too many disparate
and uncoordinated electronics recycling efforts that result in consumer
confusion—and that contribute to the likelihood that, as the group sat and
talked, another load of monitors was landing in a dumpster. While Silicon
Valley may have been the birthplace of the semiconductor, and Americans
may purchase more of its inventions than any other country, the United
States is not setting the terms of the debate when it comes to regulating
the environmental impacts of high tech.

This is true of the impacts of manufacturing as well as those of e-waste.
Any high-tech manufacturer—and the hundreds of different suppliers they
work with—that wants to sell into the global market will either have to
tailor its whole suite of products to meet the environmental regulations
of each individual jurisdiction—be it a state the size of Rhode Island or all
of Europe—or it will have to make products that are acceptable everywhere. Thus far, given the efficiencies of the high-tech industry's global
marketing and supply chain, most manufacturers seem to be choosing the
latter route and are rushing to meet hazardous materials restrictions
imposed by the European Union's RoHS directive. It's simply too cumbersome and expensive to do otherwise.

For many people globalization is synonymous with a race to the bottom. Freed to roam the globe, corporate capital can move wherever labor
and environmental regulations are weakest. Even the threat of such a
move can stifle demands for higher standards, depressing wages and weak-
ening environmental standards everywhere. Yet by making the RoHS standards essentially world standards, the high-tech industry has the opportunity to demonstrate that globalization can mean higher standards
everywhere. Similarly, the increased demand for transparency in business
practices has the potential to improve environmental standards throughout the high-tech industry's supply chain. All major high-tech manufacturers release reports—usually annually—accounting for their use of
resources and release of hazardous waste, and they do so for their facilities worldwide.

Before the antiglobalization activists accuse me of being a Pollyanna, I will say that the high-tech industry's highly proprietary and competitive nature makes it difficult for outside observers to assess thoroughly the full impacts of the materials and resources that go into high tech's products. Some measurable improvements occur by simply decreasing overall production or because reporting standards and metrics change. Still, increased public scrutiny and awareness of high environmental standards is ultimately a good thing whether you live and work in southern China, San Jose, California, or Endicott, New York. And while most people probably don't choose their high-tech equipment based on environmental profile, no company can afford to be seen as a bad actor. Many high-tech companies that have roots in Silicon Valley also have Superfund sites in their history, and they don't want to repeat that experience.

At the same time—as evidenced by the burst of state e-waste legislation—local governments and corporate-scale consumers don't want to be liable for the hazards posed by putting old high-tech equipment in landfills and incinerators, or for shouldering the entire expense of getting high-tech trash to a reputable recycler. And when enough U.S. owners of Macintosh computers find out that Apple allows its customers in Japan and Taiwan to return old equipment to the company for recycling free of charge, they may begin to ask for comparable service at home.

"I don't want to overemphasize the good," Ted Smith of the Silicon Valley Toxics Coalition said to me of Europe's RoHS and WEEE legislation, but "the two directives taken together are having a tremendous impact in harmonizing things upward."[13]

The idea of stopping and thinking ahead to the end of a product's useful life or to the long-term impacts of a newly synthesized chemical often seems antithetical to the American culture of putting the bottom line first, producing more and more, and seeking out convenience, speed, and the next new thing. The environmental changes being demanded of the high-tech industry are challenging the very American notion of succeeding by getting there first.

"When it comes to creating wealth, speed saves. It saves time, and time is money, and money is jobs, prosperity and national success," said Mitchell E. Daniels Jr. in May 2003, when he was director of the White House's Office of Management and Budget. "In almost every way, American enterprise moves more quickly than its international counterparts. While the European Union busies itself writing layer upon layer of rules, Americans are starting 600,000 new businesses every year . . . Americans fix problems as they arise; Europeans often seem bent on preventing any chance of trouble arising in the first place," said Daniels.[14]

Despite the rhetoric and pressure from the federal government and industry, when it comes to protecting human health and the environment, local governments are beginning to see the virtues of easing up on the metaphorical gas pedal. When chemicals known to interfere with thyroid hormone function and nervous system development in animals are turning up in American toddlers, nursing mothers, and young people across the United States in what scientists call record high levels,[15] many state legislators have felt compelled to act. And when potentially carcinogenic chemicals spilled a quarter-century ago continue to permeate an entire community's water supply and are seeping into homes and undermining property values, otherwise politically uninvolved citizens have begun to demand action from their elected officials, especially when faced with what often seems like interminable delay or lack of action at the federal level.

Like the demands for producer responsibility for e-waste, a shift toward precautionary policy is beginning in the United States on a local level. Where high-tech electronics are concerned, attention has focused on PBDEs, but research on other flame retardants used in high-tech electronics—tetrabromobisphenol A (TBBPA) and hexabromocyclododecane (HBCD)—indicates that these products may not be as stable or as environmentally benign as previously thought. Components of HBCD, the world's third most widely used brominated flame retardant, have been turning up in dolphins and porpoises, and in fish in the Great Lakes. Like PBDEs, HBCD has the potential to disrupt thyroid function and may also be toxic to the nervous system. Other research has shown that even very small amounts of endocrine disrupting chemicals—the thyroid is an endocrine gland—can impair the body's ability to regulate glucose and can result in obesity and diabetes.[16]

"This compound frankly scares the heck out of me in terms of the fact that it is clearly persistent and it is bioaccumulative," Linda Birnbaum, director of the EPA's experimental toxicology division told *Environmental Science and Technology* in early 2005.[17]

Also disturbed are the citizens of Endicott, New York, and other communities, who continue to wonder about the safety of their air and tap water that has been contaminated by trichloroethylene (TCE). As the summer of 2005 ended, the EPA had still not finalized the risk assessment based on its 2001 finding that TCE is sixty-five times more toxic than previously thought. Until this is completed, the EPA can't set new safety standards—including limits for TCE vapor in residential homes. In September 2005, the New York State Department of Health released a study that found excessive rates of certain cancers, birth defects, and low birth rates among Endicott residents who lived in an area known to be chemically contaminated for twenty years. No matter how TCE and other toxics are treated from now on, it will be decades before the TCE in Endicott's groundwater—or that in Mountain View, California—is gone. So the residents of Endicott and others living in communities sitting atop toxic solvents leaked from high-tech manufacturing facilities continue to wonder how these chemicals may be affecting their health.

"When one is concerned with the mysterious and wonderful functioning of the human body, cause and effect are seldom simple and easily demonstrated relationships. They may be widely separated both in place and time. To discover the agent of disease and death depends on a patient piecing together of many seemingly distinct and unrelated facts developed through a vast amount of research in widely separated fields," wrote Rachel Carson in 1962.[18] The IBM workers in upstate New York know this all too well.

Virtually all conventionally manufactured products contain synthetic chemicals, and whether or not a particular company knowingly allowed workers to be exposed to hazardous chemicals will undoubtedly be the subject of debate and litigation for years to come. But unless we change our system of rushing to use new compounds and only later assessing their impacts—and do so with participation of the influential and chemical-intensive high-tech industry—we will continue to repeat the mistakes lamented by Carson in *Silent Spring*.

Many in the business community and government believe that regulation is inherently unproductive and generally prefer a voluntary approach toward improved environmental practices. Regulation, they feel, impedes innovation, inhibits economic growth, and creates antagonism between industry and environmental advocates. Yet without regulation it's unlikely that the high-tech industry—or most others—would have made as many environmental improvements as it has since the 1970s and '80s. In many cases the industry has changed practices in advance of regulation, but without that impetus the use of numerous hazardous chemicals—be they solvents, photoresist compounds, heavy metals, or the CFCs and PFCs used in semiconductor production—would probably not have been curtailed or discontinued when they were. Industries under the spotlight for using toxic substances often say that the available alternatives may be no better or worse than what's currently employed. But that seems to me a failure of imagination.

The research and development of new, environmentally benign materials, manufacturing processes, and designs is often expensive. Companies often cite financial costs and the subsequent burden to employees and stockholders as a barrier to making such improvements. Under economic regulations that oblige publicly traded companies to maximize shareholder value—and to do so on a quarterly basis—it's hard to envision how relying solely on voluntary measures (without regulatory or financial incentives) will bring about the kind of substantial changes needed to make the high-tech industry ecologically sustainable for the long haul.

▌▌▐▌▌▐ ▌

High-tech electronics have created "virtual worlds" and foster the illusion that we have left the material world behind. Since the 1960s, as technology zoomed ahead, as a society we paid attention to what the gigabytes could bring us and ignored the mountain of plastics, metals, leaded glass, and chemicals that grew each time we upgraded our hardware.

Our embrace of the disposable has contributed to the proliferation of waste and to making consumers and taxpayers, rather than manufacturers, responsible for that waste. The Information Age has linked the world's information resources but its refuse binds us together as well. If I send a computer out for recycling, its disassembled parts will likely travel the globe as widely as did its original components. And the materials let loose as waste in manufacturing, released from discarded equipment or by the plastics of finished products, are traveling unseen through our watersheds and wafting invisibly on the breezes now coming through my open windows.

At every conference and public meeting on electronics and the environment I have attended, and in most of the conversations I've had with high-tech manufacturers, someone has mentioned the need for "a level playing field." It came up so often I began to think of this phrase as a kind of mantra. Most often used in response to proposed regulation requiring recycling and producer responsibility, the "level playing field" has become a kind of code for "no unfair advantages." No company wants to shoulder more than what it perceives as its share of responsibility, a stance impossible to argue with. But it also occurs to me that the level playing field could instead be taken to mean that we're all in this together—the ecologically interdependent individuals of Aldo Leopold's essay, "The Land Ethic." Leopold called this ethic "an evolutionary possibility and an ecological necessity," in which—to avert continued environmental degradation—we acknowledge our interdependence with the natural world.[19]

In a number of ways the high-tech industry is being asked to go where no industry has gone on such a scale before. Policies based on producer responsibility have been applied to packaging, cars, and other products in Europe and Japan, but no industry sells products into a global market the way high tech does. And none sell quite as many products in such quantity, or products of such material and design complexity, as does high tech.

If not addressed comprehensively, the problems of accumulating hazardous materials generated by and used in high-tech manufacturing and e-waste risk undermining the ecological sustainability of affected communities worldwide. Without ecological sustainability—reliable safe drinking water and food, clean air, safe and healthy working conditions, long-term health and biodiversity of fully functioning ecosystems—communities slip into disadvantaged, impoverished, unhealthy, and socially unstable conditions.

The high-tech industry has brought the reality of the global supply chain home to anyone with access to a computer and the Internet. Look at these supply-chain maps—with arrows showing the worldwide flow of materials, parts, and finished products reaching across oceans and continents—and think about how what you do with your old computer may affect the air a child breathes in Guiyu, how the quest for ores and metals affects communities in Africa and South America, and how past industrial practices are determining the future of a family in upstate New York or how current practices will determine whether it's safe to eat the fish caught in the Pacific Ocean.

We can't take back the millions of tons of high-tech trash already in the world's landfills, but we can apply some of the ingenuity that has created the products of the Digital Age to making those products ecologically sound. It may be too late to recapture the millions of gallons of leaked TCE, the perfluorocarbons lingering in the atmosphere, or the endocrine-disrupting chemicals working their way through the food web, but we can push for better-designed products and policies that should help prevent a future where we compound such problems. We may have to spend a little more for our laptops and cell phones, but that price increase would be but a tiny fraction of what it costs to clean an aquifer and acres of contaminated soil. Think of it as an insurance policy against future persistent and bioaccumulative toxic pollution; this seems to me well worth the expense.

Imagine what it would be like if upgrading software meant not having to buy a whole new computer, but simply snapping in a new processor. Or if printers and other accessories were universally compatible. Imagine what it would be like if the price of a new laptop or cell phone covered the cost of a convenient system to collect old equipment for reuse or recycling. Imagine if that price guaranteed a living wage in safe conditions to those engaged in every step of electronics disassembly, materials recovery, and manufacture. Imagine if there was no such thing as garbage.

Over fifty years ago, Aldo Leopold called for a land ethic that would change the role of the human species from conqueror to citizen of the natural world. About a decade later, when the first commercially produced computers were beginning to roll off assembly lines, Rachel Carson wrote of the arrogance of our "control of nature." Two generations have since grown up in a world whose business, education, government, scientific research, and civil society rely on high technology. In many ways high tech has divorced us from nature and has made us a bit more arrogant about our ability to overcome obstacles imposed by landscape. "Command" and "Control" are part of the everyday Information Age lexicon.

As I rattle the keyboard the wind is rustling through the maple trees in my backyard and I'm contemplating a trip up the coast. Working on this book has made me think about the curious juxtaposition of the high technology that makes my working life possible and the natural world I love to explore. I find myself thinking about the Swedish metallurgist who is finding new ways to put old computer materials to work and who knows the best places to watch for migrating birds. The Arizona miners who, as we walked a tailings berm, identified the native plants their wives gather for homeopathic remedies. The microchip engineer who told me about fishing in the bay at Osaka while visiting his colleagues overseas. The polyglot camaraderie of activists and academics crowded around a restaurant table in an old Beijing mansion. There were almost as many countries represented as chairs at that table. Cell phones rang and chopsticks clacked as shared dishes made the rounds, and conversation ranged from the e-waste conference we were attending, to the North Sea and Mumbai, and back to the next order of *shu mei*.

There was an infectious energy at that table and a sense of beginning to tackle an enormous and confounding puzzle: the global reach of the high-tech industry's environmental footprint. That footprint is not going to disappear, but we can figure out where high-tech products fit into the ecological cycle and how to reduce—and eventually eliminate—the adverse aspects of those impacts.

How to Recycle a Computer, Cell Phone, TV, or Other Digital Device

How should you recycle a computer and what should you do with your old, often obsolete, high-tech electronics devices? What follows is a list of options, though it is by no means exhaustive. Electronics recycling is an evolving business, so prices and methods of returning electronics to manufacturers and recyclers are constantly changing—as is the nature of equipment accepted by these programs. Many of these programs include options for donating working equipment for reuse. Included below are CompuMentor, Goodwill, and the National Christina Foundation, as well as some government and nonprofit organizations as sources of information, but I have not attempted to list all the possible nonprofit donation options, as many are local and the resulting list would be enormous.

To prevent your old electronics from being melted down over a rudimentary stove in Guiyu or being tossed into a landfill in Lagos, you'll want to choose a reputable recycler. Ideally, each recycler should be asked how incoming equipment is accounted for and where it—and its components— will be sent for recycling. Reuse organizations should be asked if equipment is tested before it is passed on for donation and if the group ships only working equipment; they should also be asked, specifically, who their recipient organizations are so you can check the program's legitimacy. If the answer is "We don't know" or "We can't tell you," you may want to send your equipment elsewhere, as any reputable recycler or reuse organization should be able to answer these questions. What you want to ensure is that your equipment won't be exported to parts of the world where unsafe, environmentally unsound recycling or dumping takes place—or anywhere else that you're not comfortable with. Also, ask how the recycler or reuse organization handles data destruction: Can the recycler or reuse organization wipe the hard drive for you and provide documentation that they have done so? Or can they tell you how to do this before you let go of your equipment? (Data-destruction software is available online.)

If your equipment is going to be recycled rather than donated for reuse, among the easiest options in terms of accountability is to use a manufacturer's recycling program. Because electronics manufacturers sell internationally, many of their sites include information for recycling in numerous countries. Virtually all manufacturers' U.S. take-back programs carry fees. There are hundreds of electronics manufacturers, and many do not have their own take-back programs, but of those that do, many accept others' equipment for recycling.

Once the European Union's WEEE directive is up and running, Europeans will be able to recycle electronics through the WEEE program, and information should be available through manufacturers' Web sites, local retailers, and municipalities. The same holds true in Japan and a number of other countries.

The ReThink Program hosted by eBay has a good FAQ section and many useful links, many of which are also included on the U.S. Environmental Protection Agency's "eCycling" Web site. There are also outfits like Free Geek in Portland, Oregon, that refurbish donated equipment—in whole or part—as part of a work-training program, and that build computers out of salvaged parts, which I've included here as a source of information. CompuMentor links to a Web site called "TechSoup" that has a whole section devoted to hardware recycling, including information about data-wiping software. But know that its list of recyclers—like eBay's, EIA's, and IAER's—is not vetted to weed out recyclers who export e-scrap for unsound recycling, disreputable resale, or dumping. I have also included the Basel Action Network, whose Web site carries the list of electronics recyclers that have signed BAN's stewardship pledge, under which recyclers agree not to export e-waste or to use prison labor.

Many large retailers—Best Buy, Office Depot, and Staples among them—that sell electronics participate in take-back programs, but many of these programs are intermittent. After checking the Web site, contact your local retailers (including those not listed here) to find out about any such events and programs (particularly for small items like cartridges, cell phones, and rechargeable batteries); you may want to ask to speak to a manager, as this information is often inconsistently communicated to staff.

Ink cartridges for home-office printers are often the most confusing items to recycle, as the programs vary widely and one company may not recycle all of the ink and toner cartridges it produces. There are many non-manufacturer programs for ink-cartridge recycling, but you'll probably have to do some research to find out if your cartridges qualify.

Small digital devices seem to be the hardest items to find recycling information for. I have included the Web site for Palm, although that corporation itself has no program for recycling its handheld devices. Blackberry has no specific recycling information on its site. At the other end of the size spectrum are televisions, and for those you'll probably have to ask whether a manufacturer's program is accepting TVs, particularly because so many being retired were made by companies no longer in business. The fees may differ from those for computers, but any recycler that accepts equipment with CRTs should take televisions, although almost none of the U.S.-based take-back programs run by computer manufacturers (most of which are mail-in rather than drop-off programs) accept TVs. Call ahead to check.

Another option for electronics recycling information is the public agency that regulates garbage disposal and recycling in your region. It should be able to provide you with a list of local electronics recyclers and any local take-back events. California's electronics recycling program is up and running, and Maine and Maryland's will be soon, so I have included those Web sites, as well as some regional recycling organizations.

MANUFACTURERS

Note: If Web sites have changed, use a search engine and find recycling directions by entering the manufacturer's name and the word "recycling." None of these lists are exhaustive.

Apple, http://www.apple.com/environment/recycling/
nationalservices/us.html

Canon, http://consumer.usa.canon.com/ir/controller?act
=StandardDisplayAct&keycode=recycling_disposal_info
&fcategoryid=232&modelid=8046

Dell, http://www1.us.dell.com/content/topics/segtopic.aspx/
dell_recycling?c=us&cs=19&l=en&s=dhs

Epson, http://www.epson.com/cgi-bin/Store/News.jsp?BV
_UseBVCookie=yes

Gateway, http://www.gateway.com/about/corp_responsibility/
env_options.shtml

Hewlett-Packard, http://www.hp.com/hpinfo/globalcitizenship/
environment/recycle/

IBM, http://www.ibm.com/ibm/environment/ products/
pcrservice.shtml

Lexmark, http://www.lexmark.com/recycle/

Panasonic, http://www.panasonic.com/environmental/
default.asp

Sony, http://www.sony.net/SonyInfo/Environment/recycling/
recycle/display/index.html

Toshiba, http://www.toshibadirect.com/td/b2c/ebtext.to?page
=reuse&seg=HHO

CELL PHONES AND OTHER
HANDHELD DIGITAL DEVICES

Cingular, http://www.cingular.com/about/recycling; click on links
for "Donate a phone" and "RBRC"

Motorola, http://recycling.motorola.young-america.com/index.html

Nokia, http://www.nokiausa.com/recycle?cpid=OTC-10033

Palm, http://www.palm.com/us/support/contact/environment/
   recycle.html

Sprint, http://www.sprint.com/community/communities_across/
   spc.html

T-Mobile, http://www.t-mobile.com/company/recycling/default.asp

Verizon, http://www.verizonwireless.com/b2c/aboutUs/
   communityservice/recycleOverview.jsp; click on FAQs

GOVERNMENT, NONPROFITS, AND OTHER ORGANIZATIONS

California recycling program, http://www.erecycle.org

U.S. EPA Plug-in to eCycling, http://www.epa.gov/epaoswer/osw/
   conserve/plugin/reuse.htm

Maine, Department of Environmental Protection Electronic waste
   information, http://www.state.me.us/dep/rwm/ewaste/index.htm

Maryland, "eCycling in Maryland," http://www.mde.state.md.us
   (Follow links for Recycling and "special projects.")

Northeast Recycling Council, http://www.nerc.org (Follow links for
   Electronics and Residential recycling.)

Northwest Product Stewardship Council,
   http://www.productstewardship.net/

Basel Action Network, Pledge Recyclers,
   http://www.ban.org/pledge/Locations.html

CompuMentor, http://www.compumentor.org

Free Geek, http://www.freegeek.org

Goodwill Industries, http://www1.goodwill.org/page/guest/about/
   howweoperate/recycling

National Christina Foundation, http://www.cristina.org/

eBay ReThink Program, http://rethink.ebay.com/

Electronic Industries Alliance, http://www.eiae.org/

International Association of Electronics Recyclers,
   http://www.iaer.org/search/

RETAILERS

Best Buy, http://communications.bestbuy.com/
communityrelations/recycling.asp
CompUSA
Office Depot, http://www.officedepot.com
Staples

NOTES

1.THE UNDERSIDE OF HIGH TECH

1. Rachel Carson, *Silent Spring* (Boston: Houghton Mifflin, 1962), 17.
2. Lyndon B. Johnson, "Remarks at the Signing of a Bill Establishing the Assateague Island Seashore National Park," September 21, 1965, available through the American Presidency Project, http://www.presidency.ucsb.edu.
3. Sun Yat-sen University Anthropology Department and Greenpeace China, "Anthropology Report on the Electronic Waste Disposal Industry in Guiyu, Shantao" (Hong Kong, Beijing, and Guangzhou: Sun Yat-sen University Anthropology Department and Greenpeace China, 2004), 3–4.
4. U.S. EPA Environmental Information Management System (EIMS), "Trichloroethylene Health Risk Assessment: Synthesis and Characterization," http://oaspub.epa.gov/eims/eimsapi.dispdetail?deid=23249; and via the Center for Public Environmental Oversight, http://www.cpeo.org.
5. American Water Works Association, "Stats on Tap," http://www.awwa.org/Advocacy/pressroom/statswp5.cfm.
6. Intel Museum docent, personal communication, July 26, 2005.
7. Eric Williams, Robert U. Ayers, and Miriam Heller, "The 1.7 Kilogram Microchip: Energy and Material Use in the Production of Semiconductor Devices," *Environmental Science and Technology* 36, no. 24 (2002): 5509.
8. Semiconductor Industry Association (SIA), personal communication, July 26, 2005.

9. International Association of Electronics Recyclers (IAER), *IAER Electronics Recycling Industry Report 2003* (Albany, NY: IAER, 2003), 21.

10. Eric Most, *Calling All Cell Phones: Collection, Reuse, and Recycling Programs in the US* (New York: INFORM, 2003), 1.

11. Jim Puckett, Basel Action Network, presentation at the EPR2 conference, Washington, DC, March 12–13, 2002. Comparable information is available through the IAER, the National Safety Council, and the EPA.

12. Sheila Davis and Ted Smith, *Corporate Strategies for Electronics Recycling: A Tale of Two Systems* (San Jose, CA: Silicon Valley Toxics Coalition, 2003), 3; see also the EPA's Web site, "Plug-in to eCycling," http://www.epa.gov/epaoswer/osw/conserve/plugin/.

13. Betty K. Fishbein, *Waste in the Wireless World: The Challenge of Cell Phones* (New York: INFORM, 2002), 2.

14. K. Brigden et al., *Recycling of Electronic Wastes in China and India: Workplace and Environmental Contamination* (Exeter, UK: Greenpeace International, 2005), 3.

15. Ibid., 3.

16. Cris Prystay, "Recycling E-waste," *Wall Street Journal*, September 23, 2004, B1.

17. Persistent organic pollutants (POPs) are toxic chemicals that stay in the environment for long periods of time, resist degradation, and "biomagnify as they move up through the food chain. POPs have been linked to adverse effects on human health and animals, such as cancer, damage to the nervous system, reproductive disorders, and disruption of the immune system. Because they circulate globally via the atmosphere, oceans, and other pathways, POPs released in one part of the world can travel to regions far from their source of origin." See US EPA, "Persistent Organic Pollutants (POPs) at http://www .epa.gov/oppfead1/international/pops.htm."

18. Global Futures Foundation, *Computers, E-waste and Product Stewardship: Is California Ready for the Challenge?* Report for the U.S. EPA, Region 10, http://www.crra.com/ewaste/articles/computers.html.

19. Countries that regulate e-waste disposal include Belgium, Denmark, Italy, the Netherlands, Norway, Sweden, Switzerland, Portugal, Japan, Taiwan, and Korea, and with enactment of WEEE, the entire eu.

20. IAER, *IAER Electronics Recycling Industry Report 2003*, 25, 59.

21. John Vidal, "Poisonous Detritus of the Electronic Revolution," *Guardian*, September 21, 2004.

22. IAER, *IAER Electronics Recycling Industry Report 2003*, 9, 36; Robin Ingenthron, American Retroworks, personal communication, April 19, 2005; and based on the estimate by EPA et al. that approximately 10 percent of all e-waste is recycled annually in the United States.

23. Iza Kruszewska, "Recycling of Electronic Wastes in China and India," Greenpeace International press release, August 17, 2005.

24. Nicholas Negroponte, *Being Digital* (New York: Alfred A. Knopf, 1995), 4.

25. Bills have been introduced in both the House and Senate, but as drafted neither bill would get an e-waste recycling system up and running.

26. "HP Helps Consumers Dispose of 'E-waste' with 50 Percent Off Hardware Recycling," Hewlett-Packard press release, April 21, 2005.

27. Lyndon B. Johnson, "Remarks on Conservation at a Breakfast in Portland Saluting the Northwest-Southwest Power Transmission Intertie," September 17, 1964, available at the American Presidency Project, http://www.presidency.ucsb.edu/.

28. These numbers come from the EPA's Toxics Release Inventory (TRI) Explorer database, http://www.epa.gov/triexplorer/. See also *Under the Surface*, Willamette Riverkeeper's State of the River report from 2000, available at http://www.willamette-riverkeeper.org/.

29. Bill McKibben, *Enough: Staying Human in an Engineered Age* (New York: Times Books, 2003), 217–18, 220.

2. RAW MATERIALS

1. Silicon Valley Toxics Coalition (SVTC), *Poison PCs and Toxic TVs* (San Jose, CA: SVTC, 2002), 15.

2. Timothy Townsend, Department of Environmental Engineering Sciences, University of Florida, presentation at the E-Scrap 2003 conference, Orlando, FL, October 22, 2003.

3. Jim Puckett and Ted Smith, eds., *Exporting Harm: The High-Tech Trashing of Asia* (Seattle: Basel Action Network; San Jose, CA: SVTC, 2002), 9, which cites the WEEE directive and ATSDR (Agency for Toxic Substance and Disease Registry), http://www.atsdr.cdc.gov/toxprofiles/phs46.html.

4. Electronics Industries Alliance position paper on mercury, available at http://www.eia.org (Accessed in 2004).

5. U.S. EPA Design for the Environment, *Life-Cycle Assessment of Desktop Computer Displays: Summary of Results*, "Question 8: Can the Lead, Mercury, and Liquid Crystals in Computer Displays Pose Health Risks?" http://www.epa.gov/opptintr/dfe/pubs/comp-dic/lca-sum/ques8.pdf (accessed December 2005).

6. Thomas Frank, "American Psyche," review of *The Great Divide: Retro vs. Metro America*, by John Sperling et al., *New York Times Book Review*, November 28, 2004.

7. SIA, "Media Resources: Facts and Figures," http://www.sia-online.org

8. International Copper Study Group, *World Copper Factbook* (Lisbon: International Copper Study Group, 1999), 10.

9. Donald Bleiwas and Thomas Kelly, *Obsolete Computers, "Gold Mine," or High-Tech Trash? Resource Recovery from Recycling,* U.S. Geological Survey Fact Sheet FS-060-01, July 2001.

10. Computer Industry Almanac, Inc., press release, February 28, 2003, http://c-i-a.com/pr0203.htm.

11. IAER, *IAER Electronics Recycling Industry Report 2003,* 18–19.

12. Figures on worldwide and U.S. copper consumption are from the Copper Development Association, http://www.copper.org (accessed primarily in 2004).

13. U.S. Geological Survey (USGS), "Copper: Statistics and Information," Mineral Commodity Summaries, January 2004, available at http://minerals.usgs.gov/minerals/pubs/commodity/copper/. These imports come primarily from Canada, Chile, Mexico, and Peru.

14. International Copper Study Group, *World Copper Factbook,* 16 n. 1.

15. Quotations this chapter from Pete Faur, Phelps Dodge, are from personal communication, June and July, 2004.

16. USGS, "Copper: Statistics and Information," USGS Minerals Yearbook, 2002, by Daniel Edelstein, available at http://minerals.usgs.gov/minerals/pubs/commodity/copper/. These imports come primarily from Canada, Chile, Mexico, and Peru.

17. Ken Geremia, Copper Development Association, personal communication, June 2004; Caisa Sameulsson, Minerals and Metals Recycling Research Centre, Luleå University of Technology, personal communication during May 2004 visit to Luleå, Sweden; and International Copper Study Group, *World Copper Factbook,* 21.

18. International Copper Study Group, *World Copper Factbook,* 21.

19. Ibid.

20. USGS, "Copper: Statistics and Information," Mineral Commodity Summaries, January 2004, http://minerals.usgs.gov/minerals/pubs/commodity/copper/.

21. Ibid.

22. Ibid.

23. Ibid.

24. Mineral Policy Center, *Global Mining Snapshot* (Washington, DC: Mineral Policy Center, 2003), http://www.earthworksaction.org/publications.cfm?pubID=63.

25. Ibid.

26. Ibid.

27. Alanna Mitchell, "Canada Losing Pollution Fight, Report Shows," *Toronto Globe and Mail*, December 7, 2004. The report was issued by two environmental NGOs and is available at http://www.pollutionwatch.org.

28. Earthworks and Oxfam America, *Dirty Metals: Mining, Communities and the Environment* (Washington, DC: Earthworks; Boston and Washington, DC: Oxfam America, 2004), 4.

29. Information on sulfuric acid comes from Boston University's material data safety sheet for sulfuric acid, http://www.bu.edu/es/labsafety/ESMSDSs/MSSulfuricAcid.html#anchor891201.

30. Associated Press, "Copper Giant to Pay Fine after Birds Die at Morenci Mine," August 11, 2003, available at http://www.az13.com/news/index.php?num=502.

31. Aimee Boulanger and Alexandra Gorman, *Hardrock Mining: Risks to Community Health* (Bozeman, MT: Women's Voices for the Earth, 2004), 4.

32. Timothy Padgett and Lucien O. Chauvin, "Not Golden: In Peru, Locals Are Resisting Foreign-Owned Mining Companies, Even at the Price of Continuing Poverty." *Time*, October 27, 2003.

33. Earthworks and Oxfam America, *Dirty Metals*, 4.

34. Quotations this chapter from Theo Lehner, Bolinden, are from personal communication, May 2004.

35. World Gold Council, "About Gold," http://www.gold.org/discover/knowledge/aboutgold/industrial_uses/index.html.

36. See *Dirty Metals* and other reports by Earthworks, Oxfam America, and the Mineral Policy Center. See also Robert Repetto, *Silence Is Golden, Leaden, and Copper: Disclosure of Material Environmental Information in the Hard Rock Mining Industry*, Report Number 1 (New Haven, CT: Yale School of Forestry and Environmental Studies, 2004) and news coverage in the *Denver Post*, *New York Times*, and other newspapers of the ongoing dispute between Newmont Mining Corporation and the government of Indonesia.

37. World Gold Council, "Recycled Gold," http://www.gold.org/value/markets/supply_demand/recycled.html.

38. Richard Manning, *One Round River: The Curse of Gold and the Fight for the Big Blackfoot* (New York: Henry Holt, 1997), 140.

39. World Gold Council, "Industrial Applications: Electronic Applications," http://www.gold.org/discover/sci_indu/indust_app/electronic_apps.html.

40. Bleiwas and Kelly, *Obsolete Computers*.

41. Boliden, *Annual Report 2003* (Boliden, 2004), 13.

42. United Nations Environment Programme, Mineral Resources Forum, "Mining and the Environment," http://www.mineralresourcesforum.org, and follow link for "UNEP Mining, Minerals and the Environment Programme website."

43. Quotations this chapter from Jim Moreland, Siltronic, are from personal communication, August 26, 2004; see also USGS, "Silica: Statistics and Information," Mineral Commodity Summaries, January 2004, http://minerals.usgs.gov/minerals/pubs/commodity/silica/.

44. Siltronic, "Products: Non-Polished Wafers," http://www.siltronic.com/internet/noc/Products/Non-polished/.

45. USGS, "Silicon," Mineral Commodity Summaries, January 2005, http://minerals.usgs.gov/minerals/pubs/commodity/silicmcs.05.pdf.

46. USGS, "Silica: Statistics and Information," Mineral Commodity Summaries, January 2004, http://minerals.usgs.gov/minerals/pubs/commodity/silica/.

47. National Institute for Safety and Health, "Preventing Silicosis," October 31, 1996, http://www.cdc.gov/niosh/silfact1.html.

48. National Institute for Occupational Safety and Health (NIOSH), Worker Health Chartbook 2004, NIOSH Publication No. 2004-146, http://www.cdc.gov/niosh/docs/chartbook/, search "silicosis." The World Health Organization puts U.S. deaths from silicosis at about three hundred a year and says that "the true number is not known."

49. World Health Organization, "Silicosis," Fact Sheet No. 238, May 2000, http://www.who.int/mediacentre/factsheets/en and follow the link for "Silicosis." Between 1991 and 1995 China recorded half a million cases of silicosis, and thousands of workers have been diagnosed with silicosis in India, Vietnam, and Brazil.

50. "A silicon wafer is the purest product manufactured on a commercial scale," according to Williams, Ayers, and Heller, "The 1.7 Kilogram Microchip," 5504-10.

51. Quotations this chapter from Myron Burr, Siltronic, are from personal communication, August and September 2004.

52. Williams, Ayers, and Heller, "The 1.7 Kilogram Microchip," 5504-10.

53. All of these figures are available through the EPA's TRI Explorer database, http://www.epa.gov/triexplorer/.

54. "EPA: What is the Toxics Release Inventory (TRI) Program?" http://www.epa.gov/tri/whatis.htm.

55. Wacker-Chemie GmbH, Sustainability Report: Wacker Chemie, 1998–2002 (Munich: Wacker-Chemie, n.d.), available at http://www.wacker.com.

56. Ibid.

57. As this book was going into production Eric Williams took a new post at Carnegie Mellon, but the research cited was done while he was at United Nations University in Tokyo.

58. All information from Williams and colleagues is from Williams, Ayers, and Heller, "The 1.7 Kilogram Microchip."

59. Sources include Andrius Plepys, "The environmental impacts of electronics. Going beyond the walls of semiconductor fabs." International Symposium on Electronics and the Environment, Conference Record (Piscataway, NJ: IEEE, 2004), 159–65; N. Krishnan, S. Boyd, J. Rosales et al., "Using a Hybrid Approach to Evaluating Semiconductor Life Cycle Environmental Issues," *International Symposium on Electronics and the Environment, Conference Record* (Piscataway, NJ: IEEE, 2004), 86–90.

60. Blaine Harden, "The Dirt in the New Machine," *New York Times Magazine*, August 21, 2001, sec. 6, p. 35.

61. Larry D. Cunningham, "Tantalum Recycling in the United States in 1998," U.S. Geological Survey Circular 1196-J, U.S. Department of the Interior, U.S. Geological Survey, Reston, Virgina, 2003.

62. Cabot Corporation, "What Is a Tantalum Capacitor and What Are Its Benefits?" available at http://www.cabot-corp.com.

63. Cunningham, "Tantalum Recycling in the United States in 1998," J8.

64. Cunningham, "Tantalum Recycling in the United States in 1998"; and Cabot, "What Is a Tantalum Capacitor."

65. Cabot, "What Is a Tantalum Capacitor."

66. Organisation for Economic Co-operation and Development (OECD), "OECD Key ICT Indicators," http://www.oecd.org/document/23/0,2340,en _2649_34225_33987543_1_1_1_1,00.html; and David Diggs, Wireless Foundation, and Cellular Communications and Internet Association, presentation at the E-Scrap 2004 conference, Minneapolis, MN, October 19–20, 2004.

67. BBC News, "Congo's Coltan Rush," August 1, 2001, http://news.bbc .co.uk/hi/english/world/africa/newsid-148000/1468772.stm.

68. Dena Montague, "Stolen Goods: Coltan and Conflict in the Democratic Republic of Congo," *SAIS Review* 22, no. 1 (Winter–Spring 2002): 103–18.

69. Marc Lacey, "Beyond the Bullets and Blades," *New York Times*, March 20, 2005.

70. BBC News, "DR Congo Looters Condemned," November 20, 2001, http://news.bbc.co.uk/hi/english/world/africa/newsid-1665000/ 1665952.stm.

71. United Nations Security Council, *Report of the Panel of Experts on the Illegal Exploitation of Natural Resources and Other Forms of Wealth in the Democratic Republic of Congo*, 2002.

72. Pole Institute, "The Coltan Phenomenon," November 2002, available at http://www.pole-institute.org.

73. Alex Kirby, "Mobile Phones Fuel Gorillas Plight," BBC News Online, June 11, 2002, http://news.bbc.co.uk/2/hi/science/nature/2036217.stm.

74. Pole Institute, "The Coltan Phenomenon," November 2002, available at http://www.pole-institute.org.

75. All of these figures come from Cunningham, "Tantalum Recycling in the United States in 1998."

76. Daniel Balint-Kurti, "Tin Trade Fuels Congo War," *News24.com*, March 7, 2005, http://www.news24.com/News24/Africa/Features/0,,2-11-37-1672558,00.html.

77. Tantalum-Niobium International Study Center, "About Tantalum," http://www.tanb.org.

78. H. C. Starck press release, May 24, 2002.

79. "Motorola Position on Illegally Mined Coltan," Motorola press release, August 25, 2003.

80. RAID, *Unanswered Questions: Companies, Conflict and the Democratic Republic of Congo*.

81. Antony Squazzin, "DRC Fights for Control of Its Coltan Exports," *Business Report*, February 11, 2005, www.busrep.co.za/index.php?fArticleid=2406466.

82. Balint-Kurti, "Tin Trade Fuels Congo War," *News24.com*, March 7, 2005, http://www.news24.com/News24/Africa/Features/0,,2-11-37-1672558,00.html.

83. Cunningham, "Tantalum Recycling in the Untied States in 1998," J7.

84. Ibid., J6.

3. PRODUCING HIGH TECH: ENVIRONMENTAL IMPACTS

1. Paul Freiberger and Michael Swaine, *Fire in the Valley: The Making of the Personal Computer*, 2nd ed. (New York: McGraw-Hill, 2000), 137.

2. Display caption at Intel visitor center, Hillsboro, OR.

3. Ron White, *How Computers Work*, 7th ed. (Indianapolis, IN: Que Publishing, 2004).

4. Bay Area Air Quality Management District, Air Permit Programs, sec. 7, chap. 4, "Semiconductor Manufacturing," http://www.baaqmd.gov/pmt/handbook/s07c04pd.htm.

5. International Labour Organization (ILO), *Your Health and Safety at Work: Chemicals in the Workplace*, appendix 10, "Chemical Groups," http://www.itcilo.it/actrav/actrav-english/telearn/osh/kemi/gro.htm.

6. Semiconductor Materials and Equipment International, "Materials Market Information: Technology Overview," Wafer Fab Materials Briefs: Semiconductor Photoresist Ancillaries, Removal Materials and Semiconductor Photoresists, available at http://www.semi.org.

7. Intel, "How Chips Are Made," http://www.intel.com/education/makingchips/index.htm.

8. World Watch Institute, *Vital Signs 2002* (New York: W. W. Norton, 2002), 110.

9. ILO, *Your Health and Safety at Work: Chemicals in the Workplace,* appendix 10, "Chemical Groups," http://www.itcilo.it/actrav/actrav-english/telearn/osh/kemi/gro.htm.

10. There are various explanations of this step in the fabrication process, including in White, *How Computers Work,* 44.

11. State of Utah Department of Agriculture and Food, "Milk Survey Finds Shortages in School Lunch Program," http://ag.utah.gov/pressrel/litemilk.html.

12. Williams, Ayers, and Heller, "The 1.7 Kilogram Microchip," 5505.

13. John C. Ryan and Alan Thein Durning, *Stuff: The Secret Lives of Everyday Things* (Seattle: Northwest Environment Watch, 1997), 46.

14. World Watch Institute, *Vital Signs 2002,* 110.

15. Quotations this chapter from Eric Williams, United Nations University, personal communication September 28, 2004.

16. Both pieces of correspondence were published in *Environmental Science and Technology,* and copies were provided to me by Eric Williams.

17. Jim Arthur, Hi-Line International, personal communication, September 29, 2004.

18. Quoted in Michael McCoy, "Electronic Chemicals," *Chemical and Engineering News* 81, no. 25 (June 23, 2003): 21–26. Available at http://pubs.acs.org/cen/coverstory/8125/8125electronic.html.

19. Ibid.

20. Quoted in ibid.

21. Jay M. Dietrich, "Life Cycle Process Management for Environmentally Sound and Cost Effective Semiconductor Manufacturing," in Institute of Electrical and Electronics Engineers (IEEE), *2004 IEEE International Symposium on Electronics and the Environment, Conference Record* (Piscataway, NJ: IEEE, 2004), 168.

22. White, *How Computers Work,* 4.

23. Dietrich, "Life Cycle Process Management," 168.

24. Quotations this chapter from Ted Smith, SVTC, are from personal communication, March 2004.

25. Patrick D. Eagan, Dennis M. Hussey and Robert B. Pojasek, "A Performance Model for Driving Environmental Improvement Down the Supply Chain," in IEEE, *2002 IEEE International Symposium on Electronics and the Environment, Conference Record* (Piscataway, NJ: IEEE, 2002), 107.

26. Quotations this chapter from Gary Niekerk, Intel, are from personal communication, July 2004.

27. Intel, "Twelve Years of Transparency: A Timeline of Intel's Environmental Health and Safety Accomplishments, http://www.intel.com/intel/other/ehs/timeline/index.htm.

28. Quotations this chapter from Philip Trowbridge, AMD, personal communication August 31, 2004.

29. Advanced Micro Devices (AMD), *2003 Sustainability Progress Report* (Austin, TX: AMD, 2003), http://www.amd.com/ehs/.

30. IBM, "Environmental Protection," http://ibm.com/ibm/responsibility/world/environmental, revised February 2, 2005.

31. U.S. Department of Energy, Energy Information Agency, "Emissions of Greenhouse Gases in the United States 1999," Perfluorocarbons, http://www.eia.doe.gov/oiaf/1605/gg00rpt/other_gases.html#perf.

32. SIA, SIA Issue Backgrounders, "Environmental Management," April 11, 2005, http://www.sia-online.org/backgrounders_environment.cfm.

33. U.S. Department of Energy, Energy Information Agency, "Emissions of Greenhouse Gases in the United States 1999," Perfluorocarbons, http://www.eia.doe.gov/oiaf/1605/gg00rpt/other_gases.html#perf.

34. SIA, SIA Issue Backgrounders, "Environmental Management," April 11, 2005, http://www.sia-online.org/backgrounders_environment.cfm.

35. For information on perfluorooctane sulfonate see OECD, "Perfluorooctane Sulfonate (PFOS) and Related Chemical Products," OECD, 2004, http://www.oecd.org/document/58/0,2340,en_2649_34379_2384378_1_1_1_1,00.html; and Koichi Inoue et al., "Children's Health: Perfluorooctane Sulfonate (PFOS) and Related Perfluorinated Compounds in Human Maternal and Cord Blood Samples; Assessment of PFOS Exposure in a Susceptible Population during Pregnancy," *Environmental Health Perspectives* 112, no. 11 (August 2004), http://ehp.niehs.nih.gov/members/2004/6864/6864.html.

36. U.S. EPA, "Federal Register Environmental Documents," Perfluorooctyl Sulfonates; Proposed Significant New Use Rule, http://www.epa.gov/EPA–TOX/2000/October/Day-18/t26751.htm, published in the *Federal Register* 65, no. 202 (October 18, 2000).

37. SIA, SIA Issue Backgrounders, "Environment, Safety and Health," available at http://www.sia-online.org/backgrounders.cfm.

38. U.S. EPA, "Federal Register Environmental Documents," Perfluoroalkyl Sulfonates; Proposed Significant New Use Rule, http://www.epa.gov/EPA–TOX/ 2002/ March/Day-11/t5747.htm, published in the *Federal Register* 67, no. 47 (March 11, 2002).

39. Ibid.

40. OECD, "Perfluorooctane Sulfonate (PFOS) and Related Chemical Products," OECD, 2004, http://www.oecd.org/document/58/0,2340,en_2649_34379 _2384378_1_1_1_1,00.html.

41. Jeff Myron, "Toxic Release Inventory Reporting Requirements for the Chemical Sector," ChemAlliance, May 28, 1999, http://www.chemalliance.org/ Columns /Regulatory/052899.asp.

42. Southwest Network for Environmental and Economic Justice (SNEEJ) and Campaign for Responsible Technology, *Sacred Waters: Life-Blood of Mother Earth* (San Jose, CA: SNEEJ and Campaign for Responsible Technology, 1997), 19.

43. Williams, Ayers, and Heller, "The 1.7 Kilogram Microchip," 5504.

44. SNEEJ and Campaign for Responsible Technology, *Sacred Waters*, 75, which cites a 1995 *Albuquerque Tribune* article on the soaring rate of industrial water use.

45. Ibid., 39.

46. Quatations from Gary Niekirk and Len Drago and observations from my visit to Intel in chandler, Arizona, July 12, 2004.

4. HIGH-TECH MANUFACTURE AND HUMAN HEALTH

1. U.S. EPA Technology Innovation Program, "Contaminant Focus: Trichloroethylene (TCE)," http://clu-in.org/contaminantfocus/default.focus/ sec/ Trichloroethylene_(TCE)/cat/Overview/.

2. Agency for Toxic Substances and Disease Registry (ATSDR), "Toxicological Profile for 1,1,1-Trichloroethylene," September 2004, http://www.atsdr.cdc.gov/ toxprofiles/tp70.html; Solvent History.com, "Trichloroethylene," http://www.con-centric .net/~Rnk0228/solhist.html#TCE; and U.S. EPA, Technology Innovation Program, "Contaminant Focus," http://clu-in.org/contaminantfocus/, and fol-low link for "Trichloroethylene (TCE)."

3. ATSDR, "Toxicological Profile for Trichloroethylene," September 1997, http://www.atsdr.cdc.gov/toxprofiles/tp19.html; and U.S. EPA, "Technology Transfer Network, Air Toxics Website," Trichloroethylene, http://www .epa.gov/ttn/atw/hlthef/tri-ethy.html.

4. U.S. EPA, "Fairchild Semiconductor Corp. (South San Jose Plant), California, EPA ID# CAD097012298," Site Description and History, http://www.epa.gov/Region9/cleanup/california.html and follow link under San Jose for "Fairchild Semiconductor Corp. (South)."

5. SNEEJ and Campaign for Responsible Technology, *Sacred Waters*, 20.

6. U.S. EPA, "Superfund," Frequent Questions: What is a Superfund Site? available at http://www.epa.gov/superfund.

7. From the EPA California National Priorities List, EPA ID# CAT080034234. This and all of the following EPA documents cited in this list are available at http://www.epa.gov.

8. EPA California National Priorities List, EPA ID# CAD009212838.

9. EPA California National Priorities List, EPA ID# CAD041472341.

10. EPA California National Priorities List, EPA ID# CAD009138488.

11. EPA California National Priorities List, EPA ID# CAD00915908.

12. Mountain View, CA, Office of City Manager, "Comments on June 2004 Draft Five-Year Review Report for the Middlefield-Ellis-Whisman Superfund Study Area," July 15, 2004.

13. MEW (Middlefield-Ellis-Whisman) Moffett Issues Cleanup Forum, public meeting held August 2004 in Mountain View, CA; and U.S. EPA, "Mountain View Sites Update," January 2003. Some documents are available at U.S. EPA, "Middlefield-Ellis-Whisman (MEW) Superfund Study Area, California, EPA ID# CAD982463812," http://www.epa.gov/Region9/cleanup/california.html, and follow link under Mountain View for "Middlefield-Ellis-Whisman." Documents are also available at the Mountain View Public Library.

14. Mountain View, CA, Office of City Manager, "Comments on June 2004 Draft Five-Year Review Report for the Middlefield-Ellis-Whisman Superfund Study Area," July 15, 2004.

15. S. J. Goldberg, M. D. Lebowitz, and E. Graver, "Association of High Trichloroethylene Levels in Drinking Water and Increased Incidence of Congenital Cardiac Disease in Children," 60th Scientific Sessions of the American Heart Association, Anaheim, CA, November 16–19, 1987, *Circulation* 76, no. 4:IV–359; and Associated Press, "Solvent Linked to Heart Defects Research Focuses on Water Tainted with Common Industrial Chemical," *San Jose Mercury News*, November 19, 1987.

16. U.S. EPA EIMS, "Trichloroethylene Health Risk Assessment: Synthesis and Characterization," http://oaspub.epa.gov/eims/eimsapi.dispdetail?deid=23249; and Center for Public Environmental Oversight, http://www.cpeo.org.

17. Amanda Hawes, "SCCOSH in the Early Years: A Founder's Recollections," Spring 1998, http://www.svtc.org/resource/news_let/scohist.htm.

18. Semiconductor Industry Association "Industry Facts and Figures" available at http://www.sia-online.org.

19. Susan Q. Stranahan, "The Clean Room's Dirty Secret," *Mother Jones*, March/April 2002.

20. Ibid.

21. Jian-Nan Wang, Shih-Bin Su, and How-Ran Guo, "Urinary Tract Infection among Clean-Room Workers," *Journal of Occupational Health* 44, no. 5 (2002): 329–33.

22. Rick White, former employee at IBM's Endicott, New York, plant, personal communication, November 14, 2004.

23. California Department of Health Services, "Glycol Ethers," http://www .dhs.ca.gov/ohb/HESIS/glycols.htm.

24. SVTC, "Toxic Timeline on Glycol Ethers," http://www.svtc.org/ hu_health/getime.htm.

25. Intel, "Take a Tour of Intel Ireland," http://www.intel.com/corporate/ education/emea/eng/ireland/taps/.

26. Nobelprize.org, "The Integrated Circuit," http://nobelprize.org/physics/ educational/integrated_circuit/history/.

27. U.S. EPA, "Technology Transfer Network, Air Toxics Website," Glycol Ethers, http://www.epa.gov/ttn/atw/hlthef/glycolet.html.

28. Ibid.

29. U.S. Department of Labor, Occupational Safety and Health Administration (OSHA), "Occupational Exposure to 2-Methoxyethanol, 2-Ethoxyethanol and their Acetates (Glycol Ethers)," http://www.osha.gov/pls/oshaweb/ owadisp.show_document?p_table=FEDERAL_REGISTER&p_id=18049, published in the *Federal Register* 68, no. 250 (December 31, 2003). According to OSHA's notice in the *Federal Register*, "There is now effectively only one producer of these glycol ethers remaining in the United States, Equistar Chemicals (Exs. 64-1; 64-1-1), whose production is virtually limited to closed systems so employees have little opportunity for exposure. According to ACC, Equistar exports the bulk of the glycol ethers it produces (Ex. 64-1). The *Chemical Economics Handbook* confirms this, reporting that the four glycol ethers are no longer sold in the United States (CEH 663.5000R-S). (OSHA notes that Eastman Chemical Company also produces a small amount of 2-EE in a closed system, but only for in-house use as a site-limited intermediate in the production of another product (Ex. 64-1).

"Prior to 2001, Dow Chemical Company and Union Carbide, the largest producer of these glycol ethers, produced almost 60 percent of these glycol ethers (CEH 663.5000Q). In 2001, Dow acquired Union Carbide (Exs. 64-1; 64-1-1). Last year, Dow stopped manufacturing these glycol ethers, moving instead to producing less-toxic E-series butyl glycol ethers (e.g., EB) (Exs. 64-1; 64-1-1. CEH 663.5000Q)."

30. Ethylene Glycol Ethers Panel, "Ethylene Glycol Ethers Information Update," http://www.egep.org/epge-update.htm.

31. Bob Herbert, "IBM Families Ask, 'Why?'" New York Times, September 15, 2003.

32. William M. Bulkeley, "IBM Faces Another Toxins Case That Could Be a Bigger Challenge," Wall Street Journal, March 1, 2004; William M. Bulkeley, "IBM Settles Workers' Cancer Claims," Wall Street Journal, June 24, 2004; and Peter Aronson, "Wave of IBM Suits Reach Trial," National Law Journal, February 9, 2004.

33. Bulkeley, "IBM Faces Another Toxins Case," A6. See also Bulkeley, "IBM Settles Workers' Cancer Claims."

34. SIA press release, March 2004.

35. SIA Scientific Advisory Committee, "Cancer Risk among Wafer Fabrication Workers in the Semiconductor Industry, Executive Summary," October 15, 2001, prepared by the University of Massachusetts, Lowell.

36. Michael Blanding, "The Man Who Knew Too Much," Boston Magazine, August 2004. http://www.bostonmagazine.com/ArticleDisplay.php?id=435&print=yes.

37. SIA, "SIA Worker Health Project," http://www.sia-online.org/iss_whs.cfm.

38. Presentations at the Oregon Environmental Council Healthy Environment Forum Series, Portland, OR, by Sandra Steingraber, "Toxics and Kids: What Every Parent Should Know," November 2003; John Peterson Myers, "A New View on Toxic Chemicals: How They Impact Our Health," January 28, 2004; and by Jane Houlihan, "Pollution Gets Personal: Tracking Toxic Chemicals in Our Bodies," February 19, 2004.

39. Carol L. Reinisch, Jill A. Kreilling, and Raymond E. Stephens, "A Mixture of Environmental Contaminants Increases Camp-dependent Protein Kinase in Spisula Embryos," Environmental Toxicology and Pharmacology 19, no. 1 (January 2005): 9–18, quotation from p. 16.

40. Rachel Konrad, "IBM Endangered Health of Workers, Suit Says," Oregonian, November 6, 2003.

41. Enhesa, "2004 Global Forecast of Environmental Health and Safety: Policy and Regulatory Developments Impacting Industry," Section 9: Transparency, available at http://www.enhesa.com.

42. Congressman Maurice Hinchey, New York State Assembly Standing Committee on Environmental Conservation, public hearing, Endicott, New York, November 15, 2004.

43. New York State Department of Environmental Conservation (DEC), "Environmental Contamination and Investigations at Endicott, New York," available at http://www.dec.state.ny.us.

44. Correspondence between the NY State Department of Environmental Conservation and IBM that came into the author's possession.

45. New York State DEC, "Fact Sheet: Village of Endicott Vapor Migration Project," available at http://www.dec.state.ny.us; and correspondence between the New York State DEC and IBM that came into the author's possession.

46. New York State Assembly Standing Committee on Environmental Conservation, public hearing, Endicott, New York, November 15, 2004.

47. Congressman Maurice Hinchey, New York State Assembly Standing Committee on Environmental Conservation, public hearing, Endicott, New York, November 15, 2004.

48. "Endicott Chemical Spills," *Binghamton Press and Sun Bulletin*, http://pressconnects .com/special/endicottspill.

49. New York State Department of Health, Center for Environmental Health, "Public Health Consultation Health Statistics Review: Cancer and Birth Outcome Analysis, Endicott Area, Town of Union, Broome County, New York," August 23, 2005.

50. Edie Lau, "IBM Increases Release of Possible Carcinogen," *Binghamton Press and Sun Bulletin*, July 14, 1989.

51. From the EPA's TRI Program, http://www.epa.gov/tri.

52. Dom Yanchunas, "Property Values at Risk? Realtors Say Jury Is Still Out, But Sales Haven't Slowed," *Binghamton Press and Sun Bulletin*, October 29, 2003, http://www.pressconnects.com/special/endicottspill/stories/102903-41471.shtml.

53. Tom Wilber, "245 Take Settlement from IBM," *Binghamton Press and Sun Bulletin*, March 2, 2005, http://www.pressconnects.com/special/endicottspill/stories/030205-151605.shtml.

5. FLAME RETARDANTS: A TALE OF TOXICS

1. Hans Wolkers et al., "Congener-Specific Accumulation and Food Chain Transfer of Polybrominated Diphenyl Ethers in Two Arctic Food Chains," *Environmental Science and Technology* 38, no. 6 (2004): 1667–74.

2. Myrto Petreas, California EPA, presentation at the Environmental Finance Center Network—EPA Region 9 (EFC9) Conference and Roundtable on Brominated Flame Retardants, San Francisco, September 24–25, 2002.

3. Professor Ming H. Wong, Hong Kong Baptist University, presentation at the International Conference on Electronic Waste and Extended Producer Responsibility in China, Beijing, April 21–22, 2004.

4. Northwest Environment Watch, *Flame Retardants in the Bodies of Pacific Northwest Residents* (Seattle: Northwest Environment Watch, 2004).

5. Dr. Mariann Lloyd Smith of National Toxics Network, Australia, quoted in "Hidden Poison in Australian Homes," Greenpeace Australia Pacific media release, April 13, 2005, available at http://greenpeace.org.au (search the media release archive).

6. Douglas Fischer, "What's in You," *Inside Bay Area*, March 11, 2005.

7. Tim Christie, "Biomonitoring Reveals Chemicals within Bodies," *Eugene Register Guard*, March 22, 2004.

8. "BSEF, "Human Health Research," http://www.bsef.com/env _health /human_health/index.php?/env_health/humah_health/human_health.php

9. Thomas A. MacDonald, California EPA, presentation at the EFC9 Conference and Roundtable on Brominated Flame Retardants, San Francisco, September 24–25, 2002.

10. Kim Hooper, "The PBDEs: An Emerging Environmental Challenge and Another Reason for Breast-Milk Monitoring Programs," *Environmental Health Perspectives* 108, no. 5 (May 2000): 387, http://ehpnet1.niehs.nih.gov/docs/2000/108p387-392 /hooper-full.html.

11. Susan Landry, Albemarle Corporation, presentation at the Electronic Products Recovery and Recycling (EPR2) conference, Washington, DC, March 12–13, 2002.

12. Bromine Science and Environmental Forum (BSEF), "Bromine Is Best for Fire Safety," BSEF newsletter, November 1, 2001.

13. Ibid., 1.

14. Robert Varney, EPA, presentation at the EFC9 Conference and Roundtable on Brominated Flame Retardants, San Francisco, September 24–25, 2002.

15. BSEF, "Bromine: Frequently Asked Questions," revised ed., October 2001, 6.

16. Susan D. Landry and Raymond B. Dawson, Albemarle Corporation, "Design for the Environment: Electrical and Electronic Equipment Sustainable Fire Safety," presented at the EPR2 conference, Washington, DC, March 12–13, 2002; and Thaddeus Herrick, "As Flame Retardants Build Up in People, a Ban Is Debated," *Wall Street Journal*, October 10, 2003.

17. Landry and Dawson, "Design for the Environment," presented at the EPR2 conference, Washington, DC, March 12–13, 2002.

18. Linda S. Birnbaum and Daniele F. Staskal, "Brominated Flame Retardants: Cause for Concern?" *Environmental Health Perspectives* 112, no. 1 (January 2004).

19. BFR session of the 2002 IEEE International Symposium on Electronics and the Environment, San Francisco, May 6–9, 2002.

20. Yana Kucher and Meghan Purvis, *Body of Evidence: New Science in the Debate Over Toxic Flame Retardants and Our Health,* (San Francisco: U.S. PIRG Education Fund Environment California Research and Policy Center, 2004), chart p. 14, which cites American Chemistry Council's Brominated Flame Retardant Industry Panel for the Voluntary Children's Chemical Evaluation Program (VCCEP), *Report of the Peer Consultation Meeting on Decabromodiphenyl Ether,* organized by Toxicology Excellence for Risk Assessment, September 2003.

21. USGS, "Bromine: Statistics and Information," Mineral Commodity Summaries, for years 2002 through 2005 at http://minerals.usgs.gov/minerals/pubs/commodity/bromine.

22. Ibid.; and Mineral Information Institute, "Bromine," http://www.mii.org/Minerals/photobromine.html.

23. V. M. Thomas, J. A. Bedford, and R. J. Cicerone, "Bromine Emissions From Leaded Gasoline," *Geophysical Research Letters* 24, no. 11 (June 1, 1997): 1371–74.

24. Roskill Consulting Group, "Bromine," http://www.roskill.com/reports/bromine.

25. BSEF, "Bromine: Frequently Asked Questions," revised ed., October 2001, 18.

26. ATSDR, "ToxFAQs for Polychlorinated Biphenyls (PCBs)," February 2001, http://www.atsdr.cdc.gov/tfacts17.html.

27. Quotations this chapter from Robert Hale, Virginia Institute of Marine Science, Department of Environmental Sciences, are from personal communication, October 2002.

28. Birnbaum and Staskal, "Brominated Flame Retardants: Cause for Concern?"

29. Ibid.

30. Danish Environmental Protection Agency, "Introduction to Brominated Flame Retardants," http://www.mst.dk/udgiv/Publications/1999/87-7909-416-3/html/kap01_eng.htm.

31. ATSDR, "Production, Import/Export, Use, and Disposal," chap. 7 in *Toxicological Profile for Polybrominated Biphenyls and Polybrominated Diphenyl Ethers (PBBs and PBDEs),* http://www.atsdr.cdc.gov/toxprofiles/tp68-c7.pdf.

32. Herrick, "As Flame Retardants Build Up in People."

33. Thomas A. MacDonald, presentation at the EFC9 Conference and Roundtable on Brominated Flame Retardants, San Francisco, September 24–25, 2002.

34. USGS, "Bromine: Statistics and Information," Mineral Commodity Summaries, for years 2002 through 2005 at http://minerals.usgs.gov/minerals/pubs/commodity/bromine.

35. Birnbaum and Staskal, "Brominated Flame Retardants: Cause for Concern?"

36. Linda Birnbaum, EPA, presentation at the EFC9 Conference and Roundtable on Brominated Flame Retardants, San Francisco, September 24–25, 2002.

37. Hans Wolkers et al., "Congener-Specific Accumulation and Food Chain Transfer of Polybrominated Diphenyl Ethers."

38. Presentations at the EFC9 Conference and Roundtable on Brominated Flame Retardants, San Francisco, September 24–25, 2002.

39. Mehran Alaee and Richard J. Wenning, "The Significance of Brominated Flame Retardants in the Environment: Current Understanding, Issues and Challenges," Chemosphere 46, no. 5 (February 2002): 579–82; and Ronald A. Hites, "Polybrominated Diphenyl Ethers in the Environment and People; a Meta-Analysis of Concentrations," Environmental Science and Technology 38, no. 4 (2004).

40. Hites, "Polybrominated Diphenyl Ethers in the Environment and People."

41. Presentation at the EFC9 Conference and Roundtable on Brominated Flame Retardants, San Francisco, September 24–25, 2002.

42. Ibid.

43. ATSDR, "ToxFAQs for Polychlorinated Biphenyls (PCBs)," February 2001, http://www.atsdr.cdc.gov/tfacts17.html.

44. BSEF, "Bromine: Frequently Asked Questions," revised ed., October 2001, 15.

45. BSEF, "Human Health Research," http://www.bsef.com/env_health/human_health/index.php?/env_health/human_health/human_health.php.

46. Presentation at the EFC9 Conference and Roundtable on Brominated Flame Retardants, San Francisco, September 24–25, 2002.

47. As Dr. Schecter explained it to me, this is a question of mass balance. Mass balance measures the total materials present at the beginning of a process and what is present after that process is completed as a way of quantifying what might be released to the environment or discarded as waste. It's a term that's also used in measuring how the volume of glaciers changes over time.

48. Hans Wolkers et al., "Congener-Specific Accumulation and Food Chain Transfer of Polybrominated Diphenyl Ethers"; and Ingrid Vives et al., "Polybromodiphenyl Ether Flame Retardants in Fish from Lakes in European High Mountains and Greenland," Environmental Science and Technology 38, no. 8 (2004).

49. Hites, "Polybrominated Diphenyl Ethers in the Environment and People."

50. Myrto Petreas, presentation at the EFC9 Conference and Roundtable on Brominated Flame Retardants, San Francisco, September 24–25, 2002.

51. Jon Manchester, University of Wisconsin, personal communication, October 2002.

52. For studies with research similar to Manchester's, see F. J. Ter Schure Arnout et al., "Atmospheric Transport of Polybrominated Diphenyl Ethers and Polychlorinated Biphenyls to the Baltic Sea," *Environmental Science and Technology* 38, no. 5 (March 2004): 1282–87.

53. BSEF, "Bromine: Frequently Asked Questions," revised ed., October 2001, 22.

54. Quotation from Arnold Schecter et al., "Polybrominated Diphenyl Ethers Contamination of United States Food," *Environmental Science and Technology* 38, no. 20 (2004). Many other scientists writing in peer-reviewed journals have also described PBDEs in this fashion. ATSDR, "ToxFAQs for Polychlorinated Biphenyls (PCBs)," February 2001, http://www.atsdr.cdc.gov/tfacts17.html, describes the nature of PCBs.

55. Alee and Wenning, "The Significance of Brominated Flame Retardants in the Environment."

56. BSEF, "Human Health Research," http://www.bsef.com/env_health/human _health/index.php?/env_health/human_health/human_health.php.

57. Thomas A. MacDonald, California EPA, personal communication, October 2003.

58. Hites, "Polybrominated Diphenyl Ethers in the Environment and People."

59. Fisher, "What's in You."

60. Schecter et al., "Polybrominated Diphenyl Ethers Contamination of United States Food."

61. Ronald A. Hites et al., "Global Assessment of Polybrominated Diphenyl Ethers in Farmed and Wild Salmon," *Environmental Science and Technology* 38, no. 19 (August 2004).

62. Thomas A. MacDonald, "A Perspective on the Potential Health Risks of PBDEs," *Chemosphere* 46, no. 5 (February 2002).

63. Dr. Arnold Schecter, University of Texas School of Public Health, personal communication, September 22, 2004.

64. Ibid.

65. Linda Birnbaum, EPA, personal communication, October 7, 2004.

66. For scientific findings about PBDEs, see *Report of the Peer Consultation Meeting on Decabromodiphenyl Ether*, April 2–3, 2003, University of Cincinatti, Ohio, http://www.tera.org/peer/, and search "Decabromodiphenyl Ether" under

Meeting Reports; and a companion report on octabromodiphenyl ether submitted by the American Chemistry Council's Brominated Flame Retardant Industry Panel for the Voluntary Children's Chemical Evaluation Program (VCCEP), September 30, 2003. Reports available at http://www.tera.org/peer/VCCEP/OctaPenta/OctaPentaWelcome. A report prepared for Great Lakes Chemical—part of the same Chemistry Council program—found limited evidence of potential human health risks from PBDEs in general. See *Voluntary Children's Chemical Evaluation Program Pilot (VCCEPP) Tier 1 Assessment of the Potential Health Risks to Children Associated with Exposure to the Commercial Octabromodiphenyl Ether Product, CAS No. 32536-52-0*, prepared by ENVIRON International Corp., Emeryville, CA, for Great Lakes Chemical Corp., West Lafayette, IN, April 21, 2003. But these reports relied on previously published data that didn't include the most recent evaluations of PBDEs in household dust, food, or human bodies.

67. Environmental Health Perspectives, March 10, 2003, http://ehp.niehs.nih.gov/press/031003.html

68. BSEF: "Bromine: Frequently Asked Questions, "revised ed., October 2001, 22.

69. Hewlett-Packard, "HP Standard 011-1 General Specification for the Environment—Restricted Materials," http://www.hp.com/hpinfo/globalcitizenship/environment/pdf/gse.pdf.

70. Dell, "Dell and the Environment," Dell's Brominated Flame Retardant Positions, http://www1.us.dell.com/content/topics/global.aspx/corp/environment/en/prod_design?c=us&l=en&s=corp&~section=003 (version available April 5, 2004).

71. Apple, "Apple and the Global Environment," http://www.apple.com/environment/summary.html.

72. BSEF, "Bromine: Frequently Asked Questions," revised ed., October 2001, 8.

73. Alexander Tullo, "Resting Easier," *Chemical and Engineering News*, November 17, 2003.

74. Jane Kay, "Widely Used Flame Retardant Feared to Be a Health Hazard Found in Women's Breast Tissue, Fish—Bill Would Ban It," *San Francisco Chronicle*, July 10, 2003.

75. Don Thompson, "Davis to Sign Ban on Flame Retardants," Associated Press, August 9, 2003.

76. Joan Lowy, "Maine Moves to Ban Flame Retardant," Scripps Howard News Service, April 8, 2004, http://www.knoxstudio.com/shns/story.cfm?pk=FLAMERETARDANT-04-08-04&cat=AN.

77. Linda Birnbaum, presentation at the EFC9 Conference and Roundtable on Brominated Flame Retardants, San Francisco, September 24–25, 2002.

78. Johan Eriksson et al., "Photochemical Transformations of Tetrabromobisphenol A and Related Phenols in Water," *Chemosphere* 54, no. 1 (2004): 117–26.

79. National Institute of Health Sciences, *Tetrabromobisphenol A: Review of Toxicological Literature*, June 2002, http://ntp.niehs.nih.gov/ntp/htdocs/Chem_Background/ExSumPdf/tetrabromobisphenola.pdf.

80. OSPAR Commission, "Hazardous Substance Series: Tetrabromobisphenol-A" (OSPAR Commission, 2004), http://www.ospar.org/documents/dbase/publications/p00202_BD%20on20TBBPA.pdf.

81. Paul H. Peterman, Carl E. Orazio, and Robert Gale, "Detection of Tetrabromobisphenol A and Formation of Brominiated 13c-bisphenol A's in Commercial Drinking Water Stored in Reusable Polycarbonate Containers," USGS, Columbia Environmental Research Center, Columbia, MO, http://water.usgs.gov/owq/dwi/dist_projects/carboys.htm.

82. OSPAR Commission, "Hazardous Substance Series: Tetrabromobisphenol-A," 15.

83. Roskill Consulting Group, "Bromine," http://www.roskill.com/reports/bromine.

84. R. A. Rudel et al., "Phthalates, Alkylphenols, Pesticides, Polybrominated Diphenyl Ethers, and Other Endocrine-Disrupting Compounds in Indoor Air and Dust," *Environmental Science and Technology* 35, no. 20 (2003): 4543–53.

85. World Wildlife Fund, UK, "Contamination: Results of WWF's Biomonitoring Survey," November 2003, http://www.wwf.org.uk/biotour/chemicals/asp.

86. Joel Tickner, "Putting Precaution into Practice: Implementing the Precautionary Principle," 1, excerpted from Tickner, "A Map Toward Precautionary Decision-Making," in *Protecting Public Health and the Environment: Implementing the Precautionary Principle*, ed. C. Raffensperger and J. Tickner (Washington, DC: Island Press, 1999).

87. Sandra Steingraber, *Living Downstream* (New York: Vintage Books, 1998), 270.

88. Robert Coleman, Director General European Commission, Health and Consumer Directorate, address on: "The U.S., Europe and Precaution: A Comparative Case Study Analysis of the Management of Risk in a Complex World," January 11–12, 2002, Bruges, Belgium, 4.

89. John D. Graham, "The Role of Precaution in Risk Assessment and Management: An American's View," remarks at the conference, the U.S., Europe, Precaution and Risk Management: A Comparative Case Study Analysis of the Management of Risk in a Complex World, Brussels, January 11–12, 2002. See also John D. Graham, "The Perils of the Precautionary Principle: Lessons from the American and European Experience," remarks at the Heritage

Foundation Regulatory Forum, Washington, DC, October 20, 2003, http://www.whitehouse.gov/omb/inforeg/speeches/031020graham.pdf.

90. "Mayor Brown to Sign Landmark Environmental Code in Same Ceremony as Budget," SF Environment press release, July 20, 2003, http://sfenvironment.com/articles_pr/2003/pr/073003.htm.

91. Clean Water Action is part of Alliance for a Healthy Tomorrow, a Massachusetts coalition lobbying for this suite of bills.

92. Daryl Ditz, World Wildlife Fund, personal communication, December 2003.

93. Quoted in Paul D. Thacker, "U.S. Companies Get Nervous about EU's REACH," *Environmental Science and Technology Online*, Policy News, January 5, 2005, http://pubs.acs.org/subscribe/journals/esthag-w/2005/jan/policy/pt _nervous.html. U.S. lobbying efforts were also reported by Mark Shapiro in "New Power for 'Old Europe,'" *Nation*, December 27, 2004.

94. Joel Tickner, University of Massachusetts Lowell, Department of Work Environment, personal communication, December 2003.

95. Cynthia Luppi, Clean Water Action, personal communication, December 2003.

96. Kucher and Purvis, *Body of Evidence*.

97. U.S. EPA, "Federal Register Environmental Documents," Certain Polybrominated Diphenylethers; Proposed Significant New Use Rule, http://www.epa.gov/fedrgstr/EPA-TOX/2004/December/Day-06/ t26731.htm, published in the *Federal Register* 69, no. 223 (December 6, 2004).

6. WHEN HIGH-TECH ELECTRONICS BECOME TRASH

1. Michael Paparian, California Integrated Waste Management Board, presentation at the 2002 IEEE International Symposium on Electronics and the Environment, San Francisco, May 6–9, 2002.

2. Computer Industry Almanac, Inc., http://www.c-i-a.com; and Fishbein, *Waste in the Wireless World*, 3.

3. U.S. EPA, "Acquisition Management," "Ecycling" Government Computers Under Recycling Electronics and Asset Disposition Services, http://www.epa.gov/ oamhpod1/admin_placement/0300115/fact.htm.

4. IAER, *IAER Electronics Recycling Industry Report 2003*, 20.

5. United Nations Environment Programme, "The Great E-waste Recycling Debate," http://www.grid.unep.ch/waste/html_file/36-37_ewaste.html.

6. "E-Waste Threatens India's Silicon Valley," Xinhua News Agency, http://news.xinhuanet.com/english/2004-12/11/content_2321821.htm.

7. United Nations Environment Programme, "The Great E-waste Recycling Debate," http://www.grid.unep.ch/waste/html_file/36-37_ewaste.html; their source for these statistics is the World Bank, 2002.

8. IAER, *IAER Electronics Recycling Industry Report 2003*, 20.

9. Ibid., 21.

10. "Going Flat Out," *Asia-Pacific Perspectives*, March 2005, http://www.jijigaho.or.jp/app/0503/eng/sp01.html.

11. IAER, *IAER Electronics Recycling Industry Report 2003*, 19–20.

12. OECD, "OECD Key ICT Indicators," http://www.oecd.org/document/23/0,2340,en_2649_34225_33987543_1_1_1_1,00.html; and David Diggs, presentation at the E-Scrap 2004 conference, Minneapolis, MN, October 19–20, 2004.

13. IAER, *IAER Electronics Recycling Industry Report 2003*, 20.

14. As of this writing, Lauren Roman is executive vice president of MaSeR Corporation.

15. Intel, "Moore's Law 40th Anniversary," http://www.intel.com/research/silicon/mooreslaw/index.htm.

16. IAER, *IAER Electronics Recycling Industry Report 2005*, PowerPoint presentation with report highlights.

17. Don Cressin, National Electronics Service Dealers Association, presentation at the EPR2 conference, Washington, DC, March 12–13, 2002.

18. See IAER, *IAER Electronics Recycling Industry Report 2003*.

19. Presentation at the OEC Business and the Environment Breakfast Series, Portland, OR, April 13, 2004, confirmed in a personal communication, April 15, 2004.

20. United Nations Environment Programme, "The Great E-waste Recycling Debate," http://www.grid.unep.ch/waste/html_file/36-37_ewaste.html.

21. Michael Linck, "Computer Waste Escalates," *Sioux City Journal*, March 27, 2005.

22. IAER, *IAER Electronics Recycling Industry Report 2003*, 24.

23. Ibid.

24. SVTC, *Poison PCs and Toxic TVs*, 2.

25. EPA, "Plug-in to eCycling," http://www.epa.gov/epaoswer/osw/conserve/plugin/.

26. Ibid.

27. The 1997 figure is from a presentation by Bobby Jackson, National Safety Council, at the 2002 EPR2 conference, Washington, DC, March 12–13, 2002; the 1998 figure is from National Safety Council, *Electronic Product Recovery and Recycling Baseline Report* (Washington, DC: National Safety Council, 1999), viii; and the percentage increase in volume is from SVTC, *Poison PCs and Toxic TVs*, 8.

28. IAER, *IAER Electronics Recycling Industry Report 2005*, PowerPoint presentation with report highlights.

29. Presentation by Bobby Jackson, EPR2 conference, Washington, DC, March 12–13, 2002.

30. Iza Kruszewska, Greenpeace International and Clean Production Action, "Producer Responsibility: An Essential Driver for Waste Prevention in the EU Directives on E-waste," presented at the International Conference on Electronic Waste and Extended Producer Responsibility in China, Beijing, April 21–22, 2004.

31. U.S. EPA, "Acquisition Management," "Ecycling" Government Computers Under Recycling Electronics and Asset Disposition Services, http://www.epa.gov/oamhpod1/admin_placement/0300115/fact.htm.

32. U.S. EPA, "Acquisition Management," Recycling Electronics and Asset Disposition (READ) Services for the Office of Environmental Information (OEI), RFQ-DC-03-00115, http://www.epa.gov/oamhpod1/admin _placement/0300115.

33. Wayne Rifer, Rifer Environmental, 2002 IEEE International Symposium on Electronics and the Environment, San Francisco, May 6–9, 2002.

34. Tom Geoghegan, "Facing an E-waste Mountain," *BBC News Magazine*, December 28, 2004, http://newsvote.bbc.co.uk/mpapps/pagetools/print/news.bbc.co.uk/2/hi/uk_news/magazine/4105473.stm.

35. SVTC, *Poison PCs and Toxic TVs*, 14. The European Union's annual discards included 80,000 metric tons from Sweden in 2003, population about 9 million (see El-Kresten AB, http://www.el-kretsen.se), and 100,000 metric tons generated by 7.4 million Swiss (according to Martin Eugster, SWICO, presentation at the International Conference on Electronic Waste and Extended Producer responsibility in China, Beijing, April 21–22, 2004).

36. Sony, "Recycling: Recycling Activities in Japan," http://www.sony.net/SonyInfo/Environment/recycling/recycle/japan/index.html

37. National Safety Council, *Electronic Product Recovery and Recycling Baseline Report*, 9–10.

38. IAER, *IAER Electronics Recycling Industry Report 2003*, 15.

39. Ruediger Kuehr and Eric Williams, eds., *Computers and the Environment: Understanding and Managing Their Impacts* (Dordrecht, the Netherlands: Kluwer Academic Publishers, 2003), 172.

40. SVTC press release, January 6, 2005.

41. "Dispatches: Betty Patton, Environmental Consultant," *Grist Magazine*, September 4, 2001, http://www.grist.org/comments/dispatches/2001/09/04/patton-consultant/index.html.

42. SVTC, *Poison PCs and Toxic TVs*, 2.

43. Massachusetts Department of Environmental Protection, "Cathode Ray Tube (CRT) Reuse and Recycling," http://www.state.ma.us/dep/recycle/reduce/crtqanda.htm.

44. Electronics Products Stewardship Canada, http://www.epsc.ca.

45. IAER, *IAER Electronics Recycling Industry Report 2005*, PowerPoint presentation with report highlights.

46. SVTC, *Poison PCs and Toxic TVs*, 2.

47. Best Buy stats are from Thomas Dunne, EPA, presentation at EPA Waste Summit, Washington, DC, February 28–March 1, 2005; and from "HP and Best Buy Help North Georgia Consumers Recycle E-waste," Hewlett-Packard press release, March 22, 2004, http://www.hp.com/hpinfo/newsroom/press/2004/040322c.html.

48. E-Scrap 2004 conference, Minneapolis, MN, October 19–20, 2004.

49. Ellen Simon, "As Tech Trash Piles Up, Recycling Still in Its Infancy," Associated Press, December 25, 2004.

50. Presentation at the OEC Business and the Environment Breakfast Series, Portland, OR, April 13, 2004, confirmed in a personal communication, April 15, 2004.

51. Raymond Communications, "Durables and Recycling," http://www.raymond.com/durables/.

52. SVTC press release, January 6, 2005.

53. David Diggs, presentation at the E-Scrap 2004 conference, Minneapolis, MN, October 19–20, 2004.

54. Electronic Industries Alliance, www.eia.org.

55. 2002 IEEE International Symposium on Electronics and the Environment, San Francisco, May 6–9, 2002.

56. Jim Puckett and Ted Smith, eds., *Exporting Harm: The High-Tech Trashing of Asia* (Seattle: Basel Action Network; San Jose, CA: Silicon Valley Toxics Coalition, 2002). An accompanying video by the Basel Action Network is also available.

57. Michele Raymond, "Watching Our Waste," *green@work*, January/February 2004, http://www.greenbiz.org/news/reviews_third.cfm?NewsID=26609.

58. Carl Wilmsen, *Ted Smith: Pioneer Activist for Environmental Justice in Silicon Valley, 1967–2000* (Berkeley: University of California, Regional Oral History Office, the Regents of the University of California, 2003), 123.

59. Mark Shapiro, "New Power for 'Old Europe,'" The Nation, December 27, 2004, 14.

60. Kruszewska, "Producer Responsibility," presented at the International Conference on Electronic Waste and Extended Producer Responsibility in China, Beijing, April 21–22, 2004.

61. Ibid., n. 7.

62. Beverley Thorpe, Iza Kruszewska, and Alexandra McPherson, *Extended Producer Responsibility* (London, Montreal, and Spring Brook, NY: Clean Production Action, 2004), 9.

63. EPR2 conference, Washington, DC, March 12–13, 2002.

64. Presentation at the OEC Business and the Environment Breakfast Series, Portland, OR, April 13, 2004.

65. Paul Brown, "TV and Computer Recycling Delayed after Firms' Pleas," *Guardian*, March 25, 2005.

66. Theo Lehner, quoted in *Recycling Today*, May 23, 2003, http://www .recyclingtoday.com/news/news/asp?ID=4126/.

67. Raymond Communications, "State Recycling Laws Update," *Recycling Policy NewsBriefs*, e-mail bulletin, May 30, 2002; and summaries from *E-Scrap News*, 2002–2005.

68. National Caucus of Environmental Legislators, "Enacted and Introduced PBDE Legislation 2005 Compiled by the National Caucus of Environmental Legislators" and "Introduced Electronic Waste Legislation 2003–2004," available at http://www.ncel.net; and National Caucus of Environmental Legislators, personal communication, May 2005.

69. Raymond, "Watching Our Waste."

70. Kara Reeve, Clear Water Action, presentation at the E-Scrap 2004 conference, Minneapolis, MN, October 19–20, 2004.

71. Northeast Recycling Council, "Used Electronics Market Study—Survey Analysis" (Brattleboro, VT: Northeast Recycling Council, 2003).

72. Scott Cassell, University of Massachusetts's Product Stewardship Institute, presentation at the EPR2 conference, Washington, DC, March 12–13, 2002.

73. For a summary of the legislation, see SVTC, http://www.svtc.org/cleancc/ usinit/states/ca.htm.

74. Tim Rudnicki, attorney and government affairs advocate, presentation at the E-Scrap 2004 conference, Minneapolis, MN, October 19–20, 2004.

75. Ibid.

76. Michael Janofsky, "Groups Propose Alternative to EPA Rules on Mercury," *New York Times*, November 14, 2005.

77. John Hinck, Natural Resources Council of Maine, personal communication, October 2004; and John Hinck, presentation at the E-Scrap 2004 conference, Minneapolis, MN, October 19–20, 2004. All quotations from Hinck in this chapter are from these sources.

78. Ted Smith, personal communication, March 19, 2004; and Jim Puckett, Basel Action Network, personal communication, March 2, 2004.

7. NOT IN OUR BACKYARD:
EXPORTING ELECTRONICS WASTE

1. 2002 IEEE International Symposium on Electronics and the Environment, San Francisco, May 6–9, 2002.
2. Jim Puckett, presentation at the EPR2 conference, Washington, DC, March 12–13, 2002.
3. EPR2 conference, Washington, DC, March 12–13, 2002.
4. Professor Huo Xia, presentation at the International Conference on Electronic Waste and Extended Producer Responsibility in China, Beijing, April 21–22, 2004.
5. Sun Yat-sen University Anthropology Department and Greenpeace, *Anthropology Report*.
6. Puckett and Smith, *Exporting Harm*, 22.
7. Sun Yat-sen University Anthropology Department and Greenpeace, *Anthropology Report*, 4.
8. EPR2 conference, Washington, DC, March 12–13, 2002.
9. Jim Puckett, presentation at the EPR2 conference, Washington, DC, March 12–13, 2002.
10. Sun Yat-sen University Anthropology Department and Greenpeace, *Anthropology Report*, 5.
11. Qui Bo et al., Medical College Shantou University, "Medical Investigation of E-waste Demanufacturing Industry in Guiyu Town," presented at the International Conference on Electronic Waste and Extended Producer Responsibility in China, Beijing, April 21–22, 2004.
12. Sun Yat-sen University Anthropology Department and Greenpeace, *Anthropology Report*, 9.
13. Ibid., 3, 5.
14. Ibid., 19.
15. Puckett and Smith, *Exporting Harm*, 22.
16. Brigden et al., *Recycling of Electronic Wastes in China and India*, 4.
17. Health impact information is from Center for Disease Control, "Triphenyl Phosphate," http://www.cdc.gov/niosh/pdfs/5038.pdf; and from Brigden et al., *Recycling of Electronic Wastes in China and India*.
18. Sun Yat-sen University Anthropology Department and Greenpeace, *Anthropology Report*, 82.
19. Ibid.
20. Puckett and Smith, *Exporting Harm*, 50 n. 41.

21. Ibid., 14.

22. Jerry Powell, editor of *E-Scrap News*, personal communication, April 6, 2004.

23. Chase Electronics, http://www.chaserecycling.com/. The information as quoted was on the company Web site in 2002. Similarly worded information was on their Web site as of December 2005.

24. Quoted in Peter S. Goodman, "China Serves as Dump Site for Computers," *Washington Post*, February 24, 2003, A01.

25. Toxics Link, *Scrapping the Hi-Tech Myth: Computer Waste in India* (New Delhi: Toxics Link, 2003), 26.

26. Ibid., 14.

27. Ibid.

28. Ibid., 20.

29. Ibid., 26.

30. Ibid., 21–22.

31. Ibid., 25.

32. Karl Schoenberger, "E-waste Ignored in India," *San Jose Mercury News*, December 28, 2003.

33. Brigden et al., *Recycling of Electronic Wastes in China and India*, 4.

34. Adam Minter, "China's Huge Hunger for Scrap," *Wall Street Journal*, March 25, 2004.

35. Floyd Norris, "U.S. Tech Exports Slide, but Trash Sales Are Up," *New York Times*, in January 14, 2005.

36. Sun Yat-sen University Anthropology Department and Greenpeace, *Anthropology Report*, 17.

37. Quoted in Adam Minter, "China's Advantage," *Scrap*, March/April 2003, 86.

38. IAER, *IAER Electronics Recycling Industry Report 2003*, 13.

39. Sun Yat-sen University Anthropology Department and Greenpeace, *Anthropology Report*, 15.

40. Minter, "China's Advantage."

41. Puckett and Smith, *Exporting Harm*, 2.

42. Sun Yat-sen University Anthropology Department and Greenpeace, *Anthropology Report*, 66.

43. Naoko Tojo, presentation at the International Conference on Electronic Waste and Extended Producer Responsibility in China, Beijing, April 21–22, 2004.

44. Puckett and Smith, *Exporting Harm*, 2.

45. For details on the Basel Convention see United Nations Environment Programme, Secretariat of the Basel Convention, http://www.basel.int/.

46. Sarah Westervett, Basel Action Network, personal communication, November 2005.

47. Industry Council for Electronic Equipment Recycling (ICER), *WEEE—Green List Waste Study*, report prepared for Britain's Environment Agency, April 2004, 13.

48. Jade Lee, System Services International, presentation at the IEEE International Symposium on Electronics and the Environment, San Francisco, May 6–9, 2002.

49. ICER, *WEEE—Green List Waste Study*, 13.

50. Ibid.

51. The IMPEL-TFS Seaport Project—TFS is short for transfrontier shipments—is a working group of the European Union's Network for the Implementation and Enforcement of Environmental Law (IMPEL).

52. European Union Network for Implementation and Enforcement of Environmental Law (IMPEL), *IMPEL-TFS Seaport Project: Project Report, May 2003–June 2004* (IMPEL, June 2004).

53. Ibid., 24.

54. Newsletter of the IMPEL-TFS Seaport Project, 3rd ed., June 2004.

55. Cris Prystay, "Recycling E-waste," *Wall Street Journal*, September 23, 2004, B1.

8. THE POLITICS OF RECYCLING

1. Joan Lowy, "iPod battery Target of Environmentalists," Scripps Howard News Service, March 9, 2005.

2. Jim Lynch, *Islands in the Wastestream: Baseline Study of Noncommercial Computer Reuse in the United States* (San Francisco: CompuMentor, 2004), http://www.compumentor.org/recycle/baseline-report.

3. Eric Williams, United Nations University, presentation at the E-Scrap 2004 conference, Minneapolis, MN, October 19–20, 2004.

4. Quotations this chapter from Kevin Farnam, Hewlett-Packard, are from personal communication, April 2004.

5. Scott Pencer, Noranda Recycling, personal communication, March 2004.

6. Quotations this chapter from Mark TenBrink, Noranda Recycling, are from personal communication, February 2004.

7. IAER press release, July 7, 2003.

8. 2002 IEEE International Symposium on Electronics and the Environment, San Francisco, May 6–9, 2002.

9. Cris Prystay, "Recycling E-waste," *Wall Street Journal*, September 23, 2004, B1.

10. The issue of electronics recycling standards and certification is a contentious one and has been the subject of heated debate between recyclers, environmental advocates, and industry organizations. More information can be obtained through the Basel Action Network, CHWMEG (an association of

manufacturers "globally promoting responsible waste stewardship," see http://www.chwmeg.org), the Computer TakeBack Campaign, Institute of Scrap Recycling Industries (ISRI), the International Association of Electronics Recyclers, the Silicon Valley Toxics Coaltion, and the EPA, among others.

11. Robin Ingenthron, "Setting a Higher Standard," *Recycling Today*, June 2002, http://www.recyclingtoday.com/articles/article.asp?ID=4415&CatID-8& SubCatID=20/.

12. Quotations this chapter from Robert Houghton, Redemtech, Inc., are from personal communication, June 2005.

13. Martin Eugster, presentation at the International Conference on Electronic Waste and Extended Producer Responsibility in China, Beijing, April 21–22, 2004.

14. *Recycling Today*, "Going Green International Congress and Exposition: Checking Up on Electronics Recycling," May 23, 2003, http://www.recyclingtoday.com/news/news.asp?ID=4126.

15. MBA Polymers presentation at the E-Scrap 2004 conference, Minneapolis, MN, October 19–20, 2004.

16. David Weitzman, RRT Design and Construction, presentation at the E-Scrap 2004 conference, Minneapolis, MN, October 19–20, 2004.

17. Mike Biddle, MBA Polymers, presentation at the 2002 IEEE International Symposium on Electronics and the Environment, San Francisco, May 6–9, 2002.

18. Sony, "Recycling: Recycling Activities in Japan," http://www.sony.net/SonyInfo/Environment/recycling/recycle/japan/index.html.

19. Kuusakosi Recycling, http://www.kuusakoski.com.

20. "IT's Direct Impact in the Environment," Swedish Environmental Protection Agency press release, 2003.

21. *Recycling Today*, "Going Green International Congress and Exposition: Checking Up on Electronics Recycling," May 23, 2003, http://www.recyclingtoday.com/news/news.asp?ID=4126.

22. Scandinavian Copper Development Association, "Sweden Europe's Leading Recycler of Electronic Waste," February 9, 2002, available at http://scda.com/eng.

23. Steve Skurnac presentation at the 2002 IEEE International Symposium on Electronics and the Environment, San Francisco, May 6–9, 2002.

24. Metech International, *Environmental Report: Accountable Resource Management* (Gilroy, CA, and Worcester, MA: Metech International, 2003), http://metech–arm.com.

25. E-Scrap 2004 conference, Minneapolis, MN, October 19–20, 2004.

26. Martin Eugster, presentation at the International Conference on Electronic Waste and Extended Producer Responsibility in China, Beijing, April 21–22, 2004.

27. Panasonic, "Ideas for Life: Evolution of TV Designs," http://www.panasonic.com/environmental/ecodesign_tv.asp.

28. Dell, *Environmental Report: Dell Fiscal Year 2003 in Review* (Dell, 2003), 11.

29. Hewlett-Packard, "Design for the Environment," Eco-Labels section, http://www.hp.com/hpinfo/globalcitizenship/gcreport/products/dfe.html.

30. Sony, "Customers Suppliers OEM Suppliers Sony: Green Partner Auditing," http://www.sony.net/SonyInfo/Environment/environment/management/efficiency/qfhh7c000006gqg7-att/system.pdf.

31. Timothy Mann, IBM, personal communication, November 2003.

32. Rachel Ross, "Getting the Lead Out," *Toronto Star*, August, 30, 2004.

33. Sheila Davis, then with the Materials for the Future Foundation, presentation at the 2002 IEEE International Symposium on Electronics and the Environment, San Francisco, May 6–9, 2002. In 2005 Davis became director of the Silicon Valley Toxics Coalition.

34. Elizabeth Grossman, "Toxic Recycling," *Nation*, November 21, 2005, 20–24.

35. Information on UNICOR is from Todd Baldau, Federal Prison Industries, Inc., spokesperson, e-mail communication, April 18, 2005; and from IAER, *IAER Electronics Recycling Industry Report 2003*.

36. Federal Prison Industries, Inc./UNICOR, http://www.unicor.gov.

37. Davis and Smith, *Corporate Strategies for Electronics Recycling*.

38. Pennsylvania Department of Environmental Protection, "Recycling Technical Assistance—Project Summaries," Guidance for Establishing a Permanent Electronics Recycling Program at the Wayne Township Landfill, http://www.dep.state.pa.us/dep/deputate/airwaste/wm/RECYCLE/Tech_Rpts/Clinton3.htm (accessed April 2005).

39. See the Federal Prison Industries, Inc./UNICOR's recycling page, http://www.unicor.gov/recycling.

40. Former UNICOR computer recycling worker, anonymous personal communication, August 2005.

41. *Code of Federal Regulations*, title 29, sec. 1910.1025 Lead, available at U.S. Department of Labor, OSHA, http://www.osha.gov/pls/oshaweb/owadisp.show_document?p_table=STANDARDS&p_id=10030.

42. Levels of these substances are from Bureau of Prison 2005 documents made available to me by Public Employees for Environmental Responsibility (PEER).

43. The hazardous waste classification for lead is from the Bureau of Prison 2005 documents from PEER.

44. Environmental and occupational health scientist, anonymous personal communication, April 2005.

45. Howard Hu, Harvard School of Public Health, personal communication, July 2005.
46. Grossman, "Toxic Recycling."
47. Ibid.
48. Ibid.
49. Presentation at the 2002 IEEE International Symposium on Electronics and the Environment, San Francisco, May 6–9, 2002.
50. Federal Prison Industries, Inc./UNICOR, http://www.unicor.gov/recycling/customerlist.cfm.
51. This absence was confirmed by the EPA's Katherine Osdoba, personal communication, July 2005.
52. Craig Lorch, Total Reclaim, personal communication, June 2005.
53. "Federal Prisons Admit Toxic Exposure of Staff and Inmates," PEER news release, August 24, 2005.

9. A LAND ETHIC FOR THE DIGITAL AGE

1. Hans Magnus Enzensberger, quoted in Chris Agee, "The View from Linen Hall," *Irish Pages* 2, no. 2 (Autumn/Winter 2004), 7.
2. Carson, *Silent Spring*, 13.
3. Tom Krazit, "PC Sales Strong in 2004," IDG News Service, January 19, 2005, http://www.pcworld.com/news/article/0,aid,119347,00.asp; and U.S. Census Bureau, http://www.census.gov.
4. Computer Industry Almanac, Inc., press release, June 20, 2005.
5. Ibid., press release, March 9, 2005.
6. "Global Semiconductor Sales Strong in February," SIA press release, April 2005.
7. John Greenagel, SIA, personal communication, May 2005.
8. "Computer Sales Tipped to Grow 15 Percent in 2005," *Daily Times of Pakistan*, September 26, 2005, http://www.dailytimes.com.pk/default.asp?page=story_26-9-2005_pg6_2. PC sales were expected to grow by 30 percent in India in 2005.
9. Computer Industry Almanac, Inc., press release, March 9, 2005.
10. Dana Joel Gattuso, *Mandated Recycling of Electronics: A Lose-Lose-Lose Proposition* (Washington, DC: Competitive Enterprise Institute, 2005), 25–26.
11. "Recycling Base Seeks Sustainable Development," Xinhua News Agency, http://news.xinhuanet.com/english/2005-05/23/content_2990707.htm.
12. Ibid.
13. Ted Smith, personal communication, March 19, 2004.

14. Mitchell E. Daniels Jr., "Speed Saves: Living on the Edge Is America's Edge," Oscar C. Schmidt Memorial Lecture, Rose-Hulman Institute of Technology, Office of Management and Budget, Rose-Hulman Institute of Technology, Terre Haute, IN, May 16, 2003, available at http://www.whitehouse.gov/omb/speeches/daniels051603.html.

15. Kellyn Betts, "A New Record for PBDEs in People," *Environmental Science and Technology Online*, Science News, May 25, 2005, http://pubs.acs.org/subscribe/journals/esthag-w/2005/may/science/kb_newrecord.html.

16. "Low Doses of Endocrine-Disrupting Chemicals Impair Glucogen-Releasing Alpha Cells," *Environmental Health Perspectives* press release, May 18, 2005. The full text of the research was published in *Environmental Health Perspectives* 113, no. 8 (August 2005) and is available at http://ehp.niehs.nih.gov/docs/2005/8002/abstract.html.

17. Kellyn Betts, "More Clues to HBCD Isomer Mystery," *Environmental Science and Technology Online*, Science News, March 2, 2005, http://pubs.acs.org/subscribe/journals/esthag-w/2005/mar/science/kb_hbcd.html.

18. Carson, *Silent Spring*, 189.

19. Aldo Leopold, "The Land Ethic," in *A Sand County Almanac* (New York: Ballantine Books, 1970), 238–39.

SELECTED BIBLIOGRAPHY

Ackerman, Elise."Airflow at Issue in IBM Suit." *San Jose Mercury News*, December 16, 2003.

———. "Doctor Links Cancer Cases to IBM Plant." *San Jose Mercury News*, January 14, 2004.

———."Doubt Is Cast on Tests by IBM." *San Jose Mercury News*, December 5,

Adams, Glenn. "State Now Has Plan for Your Old Television." *Portland Press Herald*, April 23, 2004.

Agency for Toxic Substances and Disease Registry (ATSDR). "Case Studies in Environmental Medicine: Trichloroethylene (TCE) Toxicity." http://www .atsdr.cdc.gov/HEC/CSEM/tce.

———."ToxFAQs for Polychlorinated Biphenyls (PCBs)," February 2001. http://www.atsdr.cdc.gov/tfacts17.html.

Alaee, Mehran, and Richard J. Wenning. "The Significance of Brominated Flame Retardants in the Environment: Current Understanding, Issues and Challenges." *Chemosphere* 46, no. 5 (February 2002).

Advanced Micro Devices (AMD). *2003 Sustainability Progress Report*. Austin, TX: AMD, 2003. http://www.amd.com/ehs/.

Apple. "Apple and the Environment." http://www.apple.com/environment/policy/.

Arctic Monitoring and Assessment Programme (AMAP). *AMAP Assessment 2002: Persistent Organic Pollutants in the Arctic*. Oslo: AMAP, 2004.

Arnout, F. J., Ter Schure, et al. "Atmospheric Transport of Polybrominated Diphenyl Ethers and Polychlorinated Biphenyls to the Baltic Sea." *Environmental Science and Technology* 38, no. 5 (March 2004): 1282–87.

Aronson, Peter. "Wave of IBM Suits Reach Trial." *National Law Journal*, February 9, 2004.

Associated Press. "Solvent Linked to Heart Defects Research Focuses on Water Tainted with Common Industrial Chemical." *San Jose Mercury News*, November 19, 1987.

Balint-Kurti, Daniel. "Tin Trade Fuels Congo War." *News24.com*, March 7, 2005. http://www.news24.com/News24/Africa/Features/0,,2-11-37-1672558,00.html.

Bättig, Hans. "States Would Do Well to Listen to Insurers Where Chips Are Concerned." September 25, 2001. http://www.converium.com/2021.asp.

BBC News. "Congo's Coltan Rush." August 1, 2001. http://news.bbc.co.uk/hi/english/world/africa/newsid-148000/1468772.stm.

———. "DR Congo Looters Condemned." November 20, 2001. http://news.bbc.co.uk/hi/english/world/africa/newsid-1665000/1665952.stm.

———. "Miners Buried in DR Congo." January 15, 2002. http://news.bbc.co.uk/hi/english/world/africa/newsid-1761000/1761540.stm

———. "Mobile Phones 'Fuel Gorillas Plight.'" June 11, 2002. http://news.bbc.co.uk/hi/english/sci/tech/newsid-2036000/2036217.stm

———. "UN Condemns Congo 'Exploitation.'" November 20, 2001. http://news.bbc.co.uk/hi/english/business/newsid-1666000/1666751.stm

Belliveau, Michael, and Stephen Lester. *PVC—Bad News Comes in Threes: The Poison Plastic, Health Hazards and the Looming Waste Crisis*. Falls Church, VA, and Portland, ME: Center for Health, Environment and Justice, Environmental Health Strategy Center, December 7, 2004.

Betts, Kellyn. "More Clues to HBCD Isomer Mystery." *Environmental Science and Technology Online*, Science News, March 2, 2005. http://pubs.acs.org/subscribe/journals/esthag-w/2005/mar/science/kb-hbcd.html

———. "New Research Challenges Assumptions about Popular Flame Retardant." *Environmental Science and Technology Online*, Science News, November 6, 2003, http://pubs.acs.org/subscribe/journals/esthag-w/2003/nov/science/kb_flame.html.

Birnbaum, Linda S., and Daniele F. Staskal. "Brominated Flame Retardants: Cause for Concern?" *Environmental Health Perspectives* 112, no. 1 (January 2004).

Blanding, Michael. "The Man Who Knew Too Much." *Boston Magazine*, August 2004. http://www.bostonmagazine.com/ArticleDisplay.php?id=435&print=yes.

Bleiwas, Donald, and Thomas Kelly. *Obsolete Computers, "Gold Mine," or High-Tech Trash? Resource Recovery from Recycling.* U.S. Geological Survey Fact Sheet FS-060-01, July 2001.

Boliden. *Annual Report 2003.* Boliden, 2004.

Borland, John. "Judge Dismisses 50 IBM Toxics Lawsuits." *CNETnews.com*, June 23, 2004.

Boulanger, Aimee, and Alexandra Gorman. *Hardrock Mining: Risks to Community Health.* Bozeman, MT: Women's Voices for the Earth, 2004.

Brigden, K., et al. *Recycling of Electronic Wastes in China and India: Workplace and Environmental Contamination.* Exeter, UK: Greenpeace International, 2005.

Bromine Science and Environmental Forum (BSEF). "Bromine: Frequently Asked Questions." Revised ed., October 2001. And 2004 online updates at http://www.bsef.com.

Brown, Paul. "TV and Computer Recycling Delayed after Firms' Pleas." *Guardian*, March 25, 2005.

Bulkeley, William M. "IBM Faces Another Toxins Case That Could Be a Bigger Challenge." *Wall Street Journal*, March 1, 2004.

———. "IBM Settles Workers' Cancer Claims." *Wall Street Journal*, June 24, 2004.

Bustillo, Miguel. "Prison-Based Recycling Effort to End." *Los Angeles Times*, August 6, 2003.

Butt, Craig M., et al. "Spatial Distribution of Polybrominated Diphenyl Ethers in Southern Ontario as Measured in Indoor and Outdoor Window Organic Films." *Environmental Science and Technology* 38, no. 3, 2004.

California Department of Health Services. "Glycol Ethers." http://www.dhs.ca.gov/ohb/HESIS/glycols.htm.

Carbone, James. "Tantalum capacitor shortages to ease in about six months." *Purchasing Magazine Online*. November 16, 2000. Available at http://www.purchasing.com.

Carson, Rachel. *Silent Spring.* Boston: Houghton Mifflin, 1962.

Chee, H. L., and K. G. Rampal. "Relation Between Sick Leave and Selected Exposure Variables among Women Semiconductor Workers in Malaysia." *Occupational Environmental Medicine* 60, April (2003): 262–270.

Chemosphere 46, no. 5 (February 2002).

Chepesiuk, Ron. "Where the Chips Fall: Environmental Health in the Semiconductor Industry." *Environmental Health Perspectives* 107, no. 9 (September 1999).

Chittum, Samme. "In an IBM Village, Pollution Fears Taint Relations With Neighbors." *New York Times*, March 15, 2004.

Christie, Tim. "Biomonitoring Reveals Chemicals within Bodies." *Eugene Register Guard*, March 22, 2004.

Colborn, Theo, Dianne Dumanoski, and John Peterson Myers. *Our Stolen Future*. New York: Dutton, 1996.

Commission of the European Communities. "Communication from the Commission on the Precautionary Principle." Brussels, February 2, 2000.

———. "Communication from the Commission towards a Thematic Strategy on the Prevention and Recycling of Waste." Brussels, May 27, 2003.

Costner, Pat, Beverley Thorpe, and Alexandra McPherson. *Sick of Dust: Chemicals in Common Products—A Needless Health Risk in Our Homes*. Spring Brook, NY: Safer Products Project, a Project of Clean Production Action, March 2005. Available at http://www.safer-products.org.

Crowson, Philip. *Minerals Handbook 2000–2001: Statistics and Analyses of the World's Minerals Industry*. Kent, UK: Mining Journal Books, 2001.

Cunningham, Larry D. *Columbium (Niobium) Recycling in the United States in 1998*. U.S. Geological Survey Circular 1196-I, U.S. Department of the Interior, U.S. Geological Survey, Reston, Virginia, 2003.

———. "Tantalum Recycling in the United States in 1998." U.S. Geological Survey Circular 1196-J, U.S. Department of the Interior, U.S. Geological Survey, Reston, Virgina, 2003.

Davis, Sheila, and Ted Smith. *Corporate Strategies for Electronics Recycling: A Tale of Two Systems*. San Jose, CA: Silicon Valley Toxics Coalition, 2003.

Dell. "Dell and the Environment." Available at http://www.dell.com.

Earthworks and Oxfam America. *Dirty Metals: Mining, Communities and the Environment*. Washington, DC: Earthworks; Boston and Washington, DC: Oxfam America, 2004.

Economist, "Coming Soon to a Laptop Near You," June 19, 2003.

Eggen, Rik I. L., et al. "Challenges in Ecotoxicology." *Environmental Science and Technology* (February 1, 2004), 59A–64A.

Electronic Industries Alliance. "Consumer Product Mercury Information Sheet." Electronic Industries Alliance, 2004.

Eljarrat, Ethel, et al. "Occurrence and Bioavailability of Polybrominated Diphenyl Ethers and Hexabromocyclododecane in Sediment and Fish from the Cinca River, a Tributary of the Ebro River (Spain)." *Environmental Science and Technology* 38, no. 9 (2004).

Elliott, Richard C., et al. "Spontaneous Abortion in the British Semiconductor Industry, an HSE Investigation." *American Journal of Industrial Medicine* 36, no. 5 (September 1999).

Environmental Data Services (ENDS). "Illegal Hazwaste Exports Go Unchecked." ENDS
     Report 353, June 2004. http://www.endsreport.com/index.cfm?action=report
     .article_printable&articleID=12755.

―――. "Potential Uses Found for TV and Computer Monitor Glass." ENDS Report 352,
     May 2004. http://www.endsreport.com/index.cfm?action=report.article
     _printable&articleID=12666.

―――. "Basel Launches Partnership on Waste Computer Shipments." ENDS Report
     353, June 2004. http://www.endsreport.com/index.cfm?action=report.article
     _printable&articleID=12756.

―――. "Dell Bows To Shareholder Pressure on Computer Recycling in the US." ENDS
     Report 341, June 2003. http://www.endsreport.com/index.cfm?action=report
     .article_printable&articleID=10007.

Erbstoesser, Greg. "IBM Offers Payments to Residents." *Binghamton Press and Sun
     Bulletin*, September 3, 2004.

E-Scrap News. Published by Resource Recycling, Portland, OR. http://www
     .resource-recycling.com

Environment News Service (ENS). "Report Names Culprits in Central Africa's
     Dirty War." ENS, April 18, 2001.

Eriksson, Johan, et al. "Photochemical Transformations of Tetrabromobisphenol
     A and Related Phenols in Water." *Chemosphere* 54, no. 1 (2004): 117–26.

Essick, Kristi. "A Call to Arms." *Industry Standard Magazine*, June 11, 2001.

European Union. Directive 2002/95/EC of the European Parliament and of the
     Council of 27 January 2003 on the restriction of the use of certain hazardous
     substances in electrical and electronic equipment. Official Journal of the
     European Union, February 3, 2003.

―――. Directive 2002/96/EC of the European Parliament and of the Council of
     27 January 2003 on waste electrical and electronic equipment (WEEE). Official
     Journal of the European Union, February 3, 2003.

European Union Network for Implementation and Enforcement of Environmental
     Law (IMPEL). *IMPEL-TFS Project on Verification of Waste Destinations Project
     Report, October 2003–November 2004*. IMPEL, November 2004. Available at
     http://europa.eu.int/comm/environment/impel.

―――. *IMPEL-TFS Seaport Project: Project Report, June 2003–May 2004*.
     IMPEL, June 2004. http://europa.eu.int/comm/environment/impel/pdf/
     tfs-projectreport.pdf.

Farrar, N. J., et al. "Atmospheric Emissions of Polybrominated Diphenyl Ethers and
     Other Persistent Organic Pollutants during a Major Anthropogenic
     Combustion Event." *Environmental Science and Technology* 38, no. 36 (2004).

Fishbein, Bette, K. "EPR: What Does It Mean? Where Is It Headed?" *P2: Pollution Prevention Review* 8, October (1998): 43–55.

———. *Waste in the Wireless World: The Challenge of Cell Phones.* New York: INFORM, 2002.

Fischer, Douglas. "The Great Experiment." *Inside Bay Area,* March 14, 2005. http://www.insidebayarea.com/portlet/article/html/fragments/print_article.jsp?article=2600903.

———. "What's in You." *Inside Bay Area,* March 11, 2005.

Fisher, Jim. "Poison Valley." *Salon.com,* July 30, 2001. http://salon.com/tech/feature/2001/07/30/almaden1.

Flynn, Laurie J. "Trial Over Safety at IBM Now in the Hands of Jurors." *New York Times,* Feburary 26, 2004.

Fordahl, Matthew. "Intel Cracks Semicondutor Barrier." *Oregonian,* November 6, 2003.

Freiberger, Paul, and Michael Swaine. *Fire in the Valley: The Making of the Personal Computer.* 2nd ed. New York: McGraw-Hill, 2000.

Gannett News Service. "Data: IBM-Endicott Harms Ozone." *Binghamton Press and Sun Bulletin,* July 13, 1989.

Gattuso, Dana Joel. *Mandated Recycling of Electronics: A Lose-Lose-Lose Proposition.* Washington, DC: Competitive Enterprise Institute, 2005.

Geoghegan, Tom. "Facing an E-waste Mountain." *BBC News Magazine,* December 28, 2004. http://newsvote.bbc.co.uk/mpapps/pagetools/print/news.bbc.co.uk/2/hi/uk_news/magazine/4105473.stm.

Gleick, James. *What Just Happened: A Chronicle from the Information Frontier.* New York: Pantheon Books, 2002.

Goodman, Peter S. "China Serves as Dump Site for Computers." *Washington Post,* February 24, 2003.

Graham, John D. "The Perils of the Precautionary Principle: Lessons from the American and European Experience." Remarks at the Heritage Foundation Regulatory Forum, Washington DC, October 20, 2003. http://www.whitehouse.gov/omb/inforeg/speeches/031020graham.pdf.

Greenpeace and Chinese Society of Environmental Sciences. Conference proceedings of the International Conference on Electronic Waste and Extended Producer Responsibility in China." Beijing, April 21–22, 2004.

Grossman, Elizabeth. "Toxic Recycling." *The Nation,* November 21, 2005, 20–24.

Grove, Andrew. *Only the Paranoid Survive.* New York: Doubleday, 1999.

Harnden, Blaine. "The Dirt in the New Machine." *New York Times Magazine,* August 21, 2001.

Harrad, Stuart, et al. "Preliminary Assessment of U.K. Human Dietary and Inhalation Exposure to Polybrominated Diphenyl Ethers." *Environmental Science and Technology* 38, no. 8 (2004).

Hassanin, Ashraf, et al. "PBDEs in European Background Soils: Level and Factors Controlling Their Distribution." *Environmental Science and Technology* 38, no. 3 (2004).

Hawkins, Jeff, with Sandra Blakeslee. *On Intelligence*. New York: Times Books, Henry Holt, 2004.

Herbert, Bob. "Early Warnings." *New York Times*, Spetember 12, 2003.

———. "IBM Families Ask, 'Why?'" *New York Times*, September 15, 2003.

Herman, Marc-Olivier, Broederlijk Delen, and Pieter Vermaerke, eds. *Supporting the War Economy in the DRC: European Companies and the Coltan Trade*. An IPIS Report. Antwerp, Belgium, January 2002. http://www.grandslacs.net/doc/2343.pdf.

Herrick, Thaddeus. "As Flame Retardants Build Up in People, a Ban Is Debated." *Wall Street Journal*, October 10, 2003.

Hites, Ronald A. "Polybrominated Diphenyl Ethers in the Environment and People; a Meta-Analysis of Concentrations." *Environmental Science and Technology* 38, no. 4 (2004).

Hites, Ronald A., et al. "Global Assessment of Polybrominated Diphenyl Ethers in Farmed and Wild Salmon." *Environmental Science and Technology* 38, no. 19 (August 2004).

Hogue, Cheryl. "Perfluorinated Pollutant Puzzle." *Chemical and Engineering News*, August 30, 2004.

Hogye, Thomas Q. "From Desktop to Dust." *Recycling Today*, November 2004. http://www.recyclingtoday.com/articles/article.asp?ID=5392&CatID=SubCatID=/.

IBM. "2002 Corporate Responsibility Report." www.ibm.com/ibm/responsibility/.

Institute of Electrical and Electronics Engineers (IEEE). *2002 IEEE International Symposium on Electronics and the Environment, Conference Record*. Piscataway, NJ: IEEE, 2002.

———. *2003 IEEE International Symposium on Electronics and the Environment, Conference Record*. Piscataway, NJ: IEEE, 2003.

———. *2004 IEEE International Symposium on Electronics and the Environment, Conference Record*. Piscataway, NJ: IEEE, 2004.

Ingenthron, Robin. "Setting a Higher Standard." *Recycling Today*, June 2002. http://www.recyclingtoday.com/articles/article.asp?ID=4415&CatID-8&SubCatID=20/.

Intel. *Global Citizenship Report and Environmental Health and Safety Report*, 2003 and 2004 eds. Chandler, AZ: Intel Corporation, 2003 and 2005. Available at http://www.intel.com.

————. *Intel Packaging Databook*. 2004 and 2005 eds. Chandler, AZ: Intel Corporation, 2004 and 2005. Available at http://www.intel.com.

Interfax. "China to Establish Scrap Electronics Recycling Systems." *Recycling Today*, January 12, 2004. http://www.recyclingtoday.com/news/news.asp?ID=5174/.

International Association of Electronics Recyclers (IAER). *IAER Electronics Recycling Industry Report 2003*. Albany, NY: IAER, 2003.

International Copper Study Group. *World Copper Factbook*. Lisbon: International Copper Study Group, 1999.

International Labour Organization (ILO). "Your Health and Safety at Work: Chemicals in the Workplace." Available at http://www.itcilo.it.

Isenberg, Anne. "Refining Semiconductors, One Atom at a Time." *New York Times*, April 8, 2004.

Kay, Jane. "Widely Used Flame Retardant Feared to Be a Health Hazard Found in Women's Breast Tissue, Fish—Bill Would Ban It." *San Francisco Chronicle*, July 10, 2003.

Kidder, Tracy. *The Soul of a New Machine*. New York: Avon Books, 1981.

Kierkegaard, Amelie, et al. "Identification of the Flame Retardant Dec-abromodiphenyl Ethane in the Environment." *Environmental Science and Technology* 38, no. 12 (2004).

Kirsch, F. William, and Gwen P. Looby. *Waste Minimization Assessment for a Manufacturer of Printed Circuit Boards*. EPA Environmental Research Brief, U.S. Environmental Protection Agency, Research and Development Risk Reduction Engineering Laboratory, Cincinnati, Ohio, July 1991.

Konrad, Rachel. "IBM Endangered Health of Workers, Suit Says." *Oregonian*, November 6, 2003.

Kucher, Yana, and Meghan Purvis. *Body of Evidence: New Science in the Debate Over Toxic Flame Retardants and Our Health*. San Francisco: U.S. PIRG Education Fund Environment California Research and Policy Center, 2004.

Kuehr, Ruediger, and Eric Williams, eds. *Computers and the Environment: Understanding and Managing Their Impacts*. Dordrecht, the Netherlands: Kluwer Academic Publishers, 2003.

Lacey, Marc. "Beyond the Bullets and Blades." *New York Times*, March 20, 2005.

Landry, Susan D., and Raymond B. Dawson. "Design for the Environment: Electrical and Electronic Equipment Sustainable Fire Safety." Paper presented at the Electronics Products Recovery and Recycling (EPR2) conference, Washington, DC, March 12–13, 2002.

Law, Robin, et al. "Levels and Trends of Polybrominated Diphenylethers and Other Brominated Flame Retardants in Wildlife." *Environment International* 29, no. 6 (September 2003): 757–70.

Lee, Jennifer. "EPA Orders Companies to Examine Effects of Chemicals." *New York Times*, April 15, 2003.

Lee, Robert G. M., et al. "PBDEs in the Atmosphere of Three Locations in Western Europe." *Environmental Science and Technology* 38, no. 3 (2004).

Linck, Michael. "Computer Waste Escalates." *Sioux City Journal*, March 27, 2005.

Lowy, Joan. "iPod battery Target of Environmentalists." Scripps Howard News Service, March 9, 2005.

———. "Maine Moves to Ban Flame Retardant." Scripps Howard News Service, April 8, 2004. http://www.knoxstudio.com/shns/story.cfm?pk=FLAMERETARDANT-04-08-04&cat=AN.

Lynch, Jim. *Islands in the Wastestream: Baseline Study of Noncommercial Computer Reuse in the United States*. San Francisco: CompuMentor, 2004. http://www.compumentor.org/recycle/baseline-report.

Malone, Michael S. *The Valley of Heart's Delight: A Silicon Valley Notebook 1963–2001*. New York: John Wiley and Sons, 2002.

Manning, Richard. *One Round River: The Curse of Gold and the Fight for the Big Blackfoot*. New York: Henry Holt, 1997.

Marquis, Christopher. "U.S. Seeks Exemption From Pesticide Ban." *New York Times*, February 7, 2003.

Matthews, H. Scott, et al. *Disposition and End-of-Life Options for Personal Computers*. Green Design Initiative Technical Report no. 97-10. Pittsburgh: Carnegie Mellon University, 1997.

McCoy, Michael. "Electronic Chemicals." *Chemical and Engineering News* 81, no. 25 (June 23, 2003): 21–26.

McDonough, William, and Michael Braungart. *Cradle to Cradle*. New York: North Point Press, 2002.

Mesquita, Jennifer, and Joanne Grower. "Uploading New Diversion Programs: Public and Private E-waste Management Initiatives." *HazMat Magazine*, August/September 2004.

Metech International. *Environmental Report: Accountable Resource Management*. Gilroy, CA, and Worcester, MA: Metech International, 2003. http://www.metech-arm.com.

Microsoft. *2004 Global Citizenship Report*. Redmond, WA: Microsoft, 2005. http://www.microsoft.com/citizenship/default.mspx.

Mineral Policy Center. *Global Mining Snapshot*. Washington, DC: Mineral Policy Center, 2003. http://www.earthworksaction.org/publications.cfm?pubID=63.

Minnesota Office of Environmental Assistance. *Management of Waste Electronic Appliances*. St. Paul, MN: Minnesota Office of Environmental Assistance, 1995.

Minter, Adam. "China's Advantage." *Scrap*, March/April 2003.

————. "China's Huge Hunger for Scrap." *Wall Street Journal*, March 25, 2004.

Mitchell, Alanna. "Canada Losing Pollution Fight, Report Shows." *Toronto Globe and Mail*, December 7, 2004.

Monosson, Emily. "Chemical Mixtures: Considering the Evolution of Toxicology and Chemical Assessment." *Environmental Health Perspectives* 113, no. 4 (April 2005).

Montague, Dena. "Stolen Goods: Coltan and Conflict in the Democratic Republic of Congo." *SAIS Review* 22, no. 1 (Winter–Spring 2002): 103–18.

Moore, Gordon E. "Cramming More Components onto Integrated Circuits." *Electronics* 38, no. 8 (April 1965).

Morris, Steven, et al. "Distribution and Fate of HBCD and TBBPA Brominated Flame Retardants in North Sea Estuaries and Aquatic Food Webs." *Environmental Science and Technology* 38, no. 21 (2004).

Most, Eric. *Calling All Cell Phones: Collection, Reuse, and Recycling Programs in the US.* New York: INFORM, 2003.

National Institute of Health Sciences. *Tetrabromobisphenol A: Review of Toxicological Literature.* June 2002. http://ntp.niehs.nih.gov/ntp/htdocs/Chem_ Background /ExSumPdf/tetrabromobisphenola.pdf.

National Safety Council. *Electronic Product Recovery and Recycling Baseline Report.* Washington, DC: National Safety Council, 1999.

Negroponte, Nicholas. *Being Digital.* New York: Alfred A. Knopf, 1995.

New Jersey Department of Health and Senior Services. "Hazardous Substance Fact Sheet: Trichlorosilane." New Jersey Department of Health and Senior Services, June 1999.

New York State Department of Environmental Conservation (DEC). "Fact Sheet: Village of Endicott Vapor Migration Project." Available at http://www.dec.state.ny.us.

Noranda, Inc. *Annual Report 2001.* http://www.noranda.com, and follow "Investor Relations" link to "Annual Reports."

Northeast Recycling Council. *Setting Up and Operating Electronics Recycling/Reuse Programs: A Manual for Municipalities and Counties.* Brattleboro, VT: Northeast Recycling Council, 2002.

————. "Used Electronics Market Study—Survey Analysis." Brattleboro, VT: Northeast Recycling Council, 2003.

O'Reilly, Finbar. "Rush for Natural Resources Still Fuels War in Congo." Reuters Foundation, August 12, 2004.

Organisation of Economic Co-operation and Development (OECD). "Perfluorooctane Sulfonate (PFOS) and Related Chemical Products." OECD, 2004. http://www.oecd.org/document/58/0,2340,fr_2649_34375_238437 _1_1_1_1,00.html.

————. *Voluntary Approaches for Environmental Policy.* Paris: OECD, 2003.

OSPAR Commission. "Hazardous Substance Series: Tetrabromobisphenol-A." OSPAR Commission, 2004. http://www.ospar.org/documents/dbase/ publications/p00202_BD%20on20TBBPA.pdf.

Padgett, Timothy, and Lucien O. Chauvin. "Not Golden: In Peru, Locals Are Resisting Foreign-Owned Mining Companies, Even at the Price of Continuing Poverty." *Time*, October 27, 2003.

Palmer, Karen, and Margaret, Walls. *The Product Stewardship Movement.* Washington, DC: Resources for the Future, 2002.

Pellow, David Naguib, and Lisa Sun-Hee Park. *The Silicon Valley of Dreams: Environmental Injustice, Immigrant Workers, and the High-Tech Global Economy.* New York: New York University Press, 2002.

Pole Institute. "The Coltan Phenomenon." November 2002. Available at http://www.pole-institute.org.

Poletti, Therese. "Worker Health." *San Jose Mercury News*, January 18, 2004.

————. "IBM Settles Birth-Defect Lawsuit." *San Jose Mercury News*, March 2, 2004.

PricewaterhouseCoopers. "China's Impact on the Semiconductor Industry." PricewaterhouseCoopers Technology Center Publications, December 2004. Available at http://www.pwc.com.

Prystay, Cris. "Companies Market to India's Have-Littles." *Wall Street Journal*, June 5, 2003.

————. "Recycling E-Waste." *Wall Street Journal*, September 23, 2004.

————. "Singapore Firm Finds Edges In 'E-waste.'" *Wall Street Journal*, September 27, 2004.

Puckett, Jim, and Ted Smith, eds. *Exporting Harm: The High-Tech Trashing of Asia.* Seattle: Basel Action Network; San Jose, CA: Silicon Valley Toxics Coalition, 2002. An accompanying video by the Basel Action Network is also available.

Pui Fong Han and May Ong. "Risks in Semiconductor Fabrication Plants." September 10, 2001. http://www.converium.com/239.asp.

Raymond, Michele. "Watching Our Waste." *green@work*, January/February 2004. http://www.greenbiz.org/news/reviews_third.cfm?NewsID=26609.

Recycling Today, "California to Begin Collecting Electronics Recycling Fee," December 31, 2004. http://www.recyclingtoday.com/news/news.asp?ID=6960/.

————,"Office Depot Launches Free Cell Phone Recycling Program," December 21, 2004. http://www.recyclingtoday.com/news/news.asp?ID=6920/.

Regalado, Antonio, and William M. Bulkeley. "IBM Cancer Data Fuel Debate Over Publication." *Wall Street Journal*, June 24, 2004.

Reinisch, Carol L., Jill A. Kreilling, and Raymond E. Stephens. "A Mixture of Environmental Contaminants Increases Camp-dependent Protein Kinase in

Spisula Embryos." *Environmental Toxicology and Pharmacology* 19, no. 1 (January 2005): 9–18.

Renner, Rebecca. "Redrawing the Dose-Response Curve." *Environmental Science and Technology*, March 1, 2004, 90A–95A.

Renout, Frank. "Europe Cracks Down on Illegal Exports of Toxic Trash." *Christian Science Monitor*, October 25, 2004. http://www.csmonitor.com/2004/1025/p07s01-woeu.html.

Repetto, Robert. *Silence Is Golden, Leaden, and Copper: Disclosure of Material Environmental Information in the Hard Rock Mining Industry.* Report Number 1. New Haven, CT: Yale School of Forestry and Environmental Studies, 2004.

Revkin, Andrew. "Bush Administration to Seek Exemptions to 2005 Ban of Pesticide."*New York Times*, January 30, 2003.

Reuters. "Court to Decide IBM Cancer Suit." *Wired News*, September 21, 2003. http://www.wired.com/news/medtech/0,1286,60528,00.html.

———. "IBM: Keep Death Records Out." *Wired News*, October 5, 2003. http://www.wired.com/news/business/0,1367,60709,00.html.

Ronen, Zeev, and Aharon Abelovich. "Anaerobic-Aerobic Process for Microbial Degradation of Tetrabromobisphenol A." *Applied and Environmental Microbiology* 66, no. 6 (June 2000).

Rubin, Leonard, and John Poate. "Ion Implantation in Silicon Technology." *Industrial Physicist*, June/July 2003.

Ryan, John C., and Alan Thein Durning. *Stuff: The Secret Lives of Everyday Things.* Seattle: Northwest Environment Watch, 1997.

Schecter, Arnold, et al. "Polybrominated Diphenyl Ethers Contamination of United States Food." *Environmental Science and Technology* 38, no. 20 (2004).

Schoenberger, Karl. "Where Computers Go to Die." *San Jose Mercury News*, November 23, 2002.

———. "Cheap Products' Human Cost." *San Jose Mercury News*, November 24, 2002.

———. "Labor Laws Enter Debate." *San Jose Mercury News*, November 25, 2002.

———. "E-waste Ignored in India." *San Jose Mercury News*, December 28, 2003.

Semiconductor Equipment Materials International. "Semiconductor Photoresists." Semiconductor Equipment Materials International, March 2003.Updated annually and available at http://www.semi.org.

SemiConFarEast.Com. "Semiconductor Manufacturing Materials." http://semiconfareast.com/semicon-matls.htm, 2004.

Semiconductor Industry Association (SIA). SIA Issue Backgrounders and press releases. http://www.sia-online.org/backgrounders.cfm.

————. "SIA Worker Health Project Press Kit 2003." San Jose, CA: SIA, 2003. Available at http://www.sia-online.org.

Shapiro, Mark. "New Power for 'Old Europe.'" *Nation*, December 27, 2004.

Sickinger, Ted. "Big Hopes on Big Screen." *Oregonian*, October 30, 2003.

Silicon Valley Toxics Coalition (SVTC). *From Silicon Valley to Green Silicon Island: Taiwan's Pollution and Promise in the Era of High-Tech Globalization*. San Jose, CA: SVTC, 2001.

————. *Poison PCs and Toxic TVs*. San Jose, CA: SVTC, 2002. Available at http://www.svtc.org along with a 2004 edition..

Simon, Ellen. "As Tech Trash Piles Up, Recycling Still in Its Infancy." Associated Press, December 25, 2004.

Skrzycki, Cindy. "OSHA Slow to Act on Beryllium Exposure, Critic Says." *Washington Post*, January 31, 2005. http://www.washingtonpost.com/ac2/wp-dyn/A52864-2005Jan31?language=printer.

————. "Report Sheds Light on Changing Role of Regulation." *Washington Post*, January 25, 2005. http://www.washingtonpost.com/ac2/wp-dyn/A34060-2005 Jan24?language=printer.

Squazzin, Antony. "DRC Fights for Control of Its Coltan Exports." *Business Report*, February 11, 2005. http://www.busrep.co.za/index.php?fArticleid = 2406466.

Sony. "Orange R-Net: Sony High-quality Foamed Polystyrene Recycling System." www.sony.net/SonyInfo/Environment/recycle/development/01/qfhh7c000000 54ci-att/qfhh7c00000054gw.pdf.

Southwest Network for Environmental and Economic Justice (SNEEJ) and Campaign for Responsible Technology. *Sacred Waters: Life-Blood of Mother Earth*. San Jose, CA: SNEEJ and Campaign for Responsible Technology, 1997.

Stapleton, Heather M., et al. "Debromination of Polybrominated Diphenyl Ether Congeners BDE 99 and BDE 183 in the Intestinal Tract of the Common Carp (Cyprinus carpio)." *Environmental Science and Technology* 38, no. 4 (2004).

Steingraber, Sandra. *Living Downstream*. New York: Vintage Books, 1998.

Stevels, Ab. "Experiences with the Take-Back of White and Brown Goods in the Netherlands." In the proceedings of the 2nd International Symposium on Environmentally Conscious Design and Inverse Manufacturing, Tokyo. Ecodesign, 2001.

Stevens, E. S. *Green Plastics: An Introduction to the New Science of Biodegradable Plastics*. Princeton, NJ: Princeton University Press, 2002.

Stranahan, Susan Q. "The Clean Room's Dirty Secret." *Mother Jones*, March/April 2002.

Sun Yat-sen University Anthropology Department and Greenpeace China. *Anthropology Report on the Electronic Waste Disposal Industry in Guiyu, Shantao.* Hong Kong, Beijing, and Guangzhou: Sun Yat-sen University Anthropology Department and Greenpeace China, 2004.

Taub, Eric A. "The Long Last Gasp of Tube-Based TV." *New York Times,* October 29, 2003.

TCO Development. "Computers" and "Mobile Phones." Available at http://www.tcodevelopment.com.

Tenenbaum, David J. "Short-Circuiting Environmental Protections?" *Environmental Health Perspectives* 111, no. 5 (May 2003).

Thacker, Paul D. "U.S. Companies Get Nervous about EU's REACH." *Environmental Science and Technology Online,* Policy News, January 5, 2005. http://pubs.acs.org /subscribe/journals/esthag-w/2005/jan/policy/pt_nervous.html.

Thomas, V. M., J. A. Bedford, and R. J. Cicerone. "Bromine Emissions From Leaded Gasoline." *Geophysical Research Letters* 24, no. 11 (June 1, 1997): 1371–74.

Thompson, Don. "Davis to Sign Ban on Flame Retardants," Associated Press, August 9, 2003.

Thorpe, Beverley, Iza Kruszewska, and Alexandra McPherson. *Extended Producer Responsibility.* London, Montreal, and Spring Brook, NY: Clean Production Action, 2004.

Tickner, Joel, ed. *Precaution, Environmental Science, and Preventative Public Policy.* Washington, DC: Island Press, 2003.

Tomy, Gregg T., et al. "Bioaccumulation, Biotransformation, and Biochemical Effects of Brominated Diphenyl Ethers in Juvenile Lake Trout (Savelinus manaycush)." *Environmental Science and Technology* 38, no. 5, 2004.

Toxics Link. *Scrapping the Hi-Tech Myth: Computer Waste in India.* New Delhi: Toxics Link, 2003.

Tullo, Alexander. "Resting Easier." *Chemical and Engineering News,* November 17, 2003.

Ueno, Daisuke, et al. "Global Pollution Monitoring of Polybrominated Diphenyl Ethers Using Skipjack Tuna as a Bioindicator." *Environmental Science and Technology* 38, no. 8, 2004.

U.S. Environmental Protection Agency (EPA). "Acquisition Management." "Ecycling" Government Computers Under Recycling Electronics and Asset Disposition Services. http://www.epa.gov/oamhpod1/admin_placement/0300115/fact.htm.

———. "Trichloroethylene Health Risk Assessment: Synthesis and Characterization: An EPA Science Advisory Board Report." A review by the

TCE Review Panel of the Environmental Health Committee of the U.S. EPA Science Advisory Board (SAB), December 2002, http://www.epa.gov/ sab/pdf/ehc03002.pdf.

———. Toxics Release Inventory (TRI) Explorer database. http://www.epa.gov/ triexplorer/.

U.S. Geological Survey (USGS). "Columbium (Niobium) and Tantalum." USGS Minerals Yearbook, 2000, by Larry D. Cunningham. Available at http://minerals .usgs.gov/minerals/pubs/commodity/niobium.

———."Copper: Statistics and Information." http://minerals.usgs.gov/minerals /pubs/commodity/copper/.

———."Copper: Statistics and Information." USGS Minerals Yearbook, 2002, by Daniel Edelstein. Available at http://minerals.usgs.gov/minerals/pubs/com- modity/copper/.

———."Niobium (Columbium) and Tantalum: Statistics and Infor- mation."Also Mineral Commodity Summaries for 2001–2003. Available at http://minerals.usgs.gov/minerals/pubs/commodity/niobium.

———."Silica: Statistics and Information." http://minerals.usgs.gov/minerals/ pubs/commodity/silica/.

Vidal, John. "Poisonous Detritus of the Electronic Revolution." *Guardian*, September 21, 2004.

Vives, Ingrid, et al. "Polybromodiphenyl Ether Flame Retardants in Fish from Lakes in European High Mountains and Greenland." *Environmental Science and Technology* 38, no. 8 (2004).

Vos, Joseph G., et al. "Brominated Flame Retardants and Endocrine Disruption." *Pure Applied Chemistry* 75, nos. 11–12 (2003): 2039–46.

Wacker-Chemie GmbH. *Sustainability Report: Wacker-Chemie, 1998–2002*. Munich: Wacker-Chemie GmbH, n.d. Available at http://www.wacker.com.

Wakabi, Wairagala. "Uganda Hopes Coltan Will Attract Investors." *East African*, April 29, 2002.

Wang, Jian-Nan, Shih Bin-Su, and How-Ran Guo. "Urinary Tract Infection among Clean-Room Workers." *Journal of Occupational Health* 44, no. 5 (2002): 329–33.

Williams, Eric, Robert U. Ayers, and Miriam Heller. "The 1.7 Kilogram Microchip: Energy and Material Use in the Production of Semiconductor Devices." *Environmental Science and Technology* 36, no. 24 (2002).

Wilber, Tom. "245 Take Settlement from IBM." *Binghamton Press and Sun Bulletin*, March 2, 2005. http://www.pressconnects.com/special/endicottspill/stories /030205-151605.shtml.

Wilber, Tom. "Endicott Spills: Some Lose Aid for Cleanup as Rules Change." *Binghamton Press and Sun Bulletin*, December 14, 2004. Other articles about Endicott are archived at http://www.pressconnects.com under "Endicott Spill."

White, Ron. *How Computers Work*. 7th ed. Indianapolis, IN: Que Publishing, 2004.

WHO/Convention Task Force on the Health Aspects of Air Pollution. *Health Risks of Persistent Organic Pollutants From Long-Range Transboundary Air Pollution*. World Health Organization, 2003. http://www.euro.who.int/Document/e78963.pdf.

Williams, Eric. "Environmental Impacts of Microchip and Computer Production." Presentation at E-Scrap 2004 conference, Minneapolis, MN, October 19–20, 2004.

Wilmsen, Carl. *Ted Smith: Pioneer Activist for Environmental Justice in Silicon Valley, 1967–2000*. Berkeley: University of California, Regional Oral History Office, the Regents of the University of California, 2003.

Wolkers, Hans, et al. "Congener-Specific Accumulation and Food Chain Transfer of Polybrominated Diphenyl Ethers in Two Arctic Food Chains." *Environmental Science and Technology* 38, no. 6 (2004): 1667–74.

Women's Foundation of California. "Confronting Toxic Contamination in Our Communities: Women's Health and California's Future." San Francisco: Women's Foundation of California, October 2003. Available at http://www.womensfoundca.org.

Wonacott, Peter. "Green Groups Bloom in China." *Wall Street Journal*, June 15, 2004.

World Watch Institute. *Vital Signs 2002*. New York: W. W. Norton, 2002.

World Wildlife Fund. *Bad Blood? A Survey of Chemicals in the Blood of European Ministers*. Brussels: World Wildlife Fund, October 2004. http://www.worldwildlife.org/toxics/pubs/badblood.pdf.

Zogbi, Dennis M. "The Tantalum Supply Chain 2000–2001." TTI, Inc., January 14, 2002. http://www.ttiinc.com/object/ME_zogbi20020114.html.

INDEX

Georgia, 201
Geremia, Ken, 24
German Federal Environment Agency, 122
German Packaging Ordinance, Green Dot, 160
Germany, 35–36, 42, 45, 48, 67, 160, 204, 212, 242, 254
Ghana, 201
Gilroy (California), 233, 237
Gisenyi (Rwanda), 47
glass, 20, 223, 233, 248; (types of): leaded glass, 165, 177, 187–88, 193, 224–25, 236, 256, 262
global supply chain, 45, 52, 55, 66, 243
globalization, 156, 258–59, 264
glycol ethers, 7, 56, 88–89, 91–94, 96–97
gold, xii, 2, 18, 22, 28–32, 37, 44 , 52, 59, 60, 187, 191–92, 217, 224, 231–32, 234, 236–39, 254
Gold Institute, 28
Goldberg, Stanley, 84
Goma (DRC), 47
Goodwill, 150, 153
gorillas, (types of): lowland gorilla, 47
Graham, John D., 135
GrassRoots Recycling Network, 157
Gray, George, 134
Great Lakes, 112, 260
Great Lakes Chemical Corporation, 119, 129, 132
Great Oaks Water Company, 76
Greenagel, John, 254
greenhouse gases, 41, 67, 68
Greenland, xi, 113
Greenock (Scotland), 93
Greenpeace, 116; Greenpeace China, 183, 185, 187–89, 194–95, 197, 199, 206, 208; Greenpeace International, 156, 160, 172, 193, 195, 203, 238
groundwater contamination, *See* water pollution
GTE, 79, 81
Guangdong Province, 184
Guangzhou(China), 188, 197, 257
Guiyu (China), 2, 12, 184, 186–88, 193, 194, 197–99, 206, 257, 263
Gulf of Bothnia, 224
Gulf of Mexico, 113

Hale, Robert, 120, 122–25
halogen, 115, 245; halogenated compounds, 58, 115, 245
Hamburg (Germany), 204
Harden, Blaine, 45

Harvard Center for Risk Analysis, 134
Harvard School of Public Health, 239
Hawaii, 162
Hawes, Amanda, 85–89, 92, 95, 97, 99
health impacts (*see also* worker health and safety): 7, 8, 12, 16, 58, 64–65, 75–77, 91–94, 97, 101–2, 107, 109, 119, 127, 197–98, 255, 260–61; assessment of 7, 262
Hennepin County, 177
Herbert, Bob, 94
Hernandez, Alida, 86, 96, 98
herring gull, 113
Hetch Hetchy, 81
Hewlett-Packard (HP), 41, 48, 54, 79, 128, 150–51, 159, 170, 174, 179, 214–216, 218–19, 222, 227, 240–41, 242–44, 292
hexabromocyclododecane (HBCD), 120, 260
hexamethyldisilazane, 58
high-impact polystyrene (HIPS), 117
Hillcrest (New York), 106
Hillsboro (Oregon), 60
Hilton, Bryant, 174–75
Hinchey, Maurice, 99, 104–6
Hinck, Jon, 178–189
Hitachi, 180
Hites, Ronald, 117, 121, 123–27, 130, 132
Holland, 1, 113 (*see also* The Netherlands)
Holmen (Wisconsin), 247
Honeywell, 73, 79
Hong Kong, 5, 15, 142, 156, 184, 191, 194, 204
Hong Kong Baptist University, 198
Hooper, Kim, 114, 121
Hopewell Junction (New York), 107
Houghton, Robert, 222
Houston, Shelby, 226
Houston (Texas), 216
Hu, Howard, 249
Huang Jian Zhong, 208
Hudak, Wanda, 92, 110
Huo Xia, 197
Huron Associates, 100, 109
Hutu, 47
Hydrochloric acid, 41, 57–58, 187
hydrofluoric acid, 20, 38, 59
hydrogen, 38; hydrogen chloride, 38; hydrogen fluoride, 41
hydroxyl monoethanolamine, 59

Iceland, 143
Idaho, 162
Illinois, 173
immune system, 120, 131
incinerator, 44, 141, 147, 190, 226, 255, 259; incineration, 179, 234

ABOUT THE AUTHOR

Elizabeth Grossman is the author of *Watershed: The Undamming of America*, *Adventuring Along the Lewis and Clark Trail,* and co-editor of *Shadow Cat: Encountering the American Mountain Lion.* Her work has appeared in a variety of publications including *The Nation, Orion, The Seattle Times,* and the *Washington Post.* She lives in Portland, Oregon.